MAKING
PLASTIC-LAMINATE
COUNTERTOPS

MAKING
PLASTIC-LAMINATE
COUNTERTOPS

HERRICK KIMBALL

The Taunton Press

COVER PHOTO: **SCOTT PHILLIPS**

for fellow enthusiasts

First printing: 1996

Printed in the United States of America

A FINE HOMEBUILDING Book

Fine Homebuilding® is a trademark of The Taunton Press, Inc.,
registered in the U.S. Patent and Trademark Office.

The Taunton Press, 63 South Main Street, PO Box 5506,
Newtown, CT 06470-5506

Library of Congress Cataloging-in-Publication Data

Kimball, Herrick.
    Making plastic-laminate countertops / Herrick Kimball.
        p.   cm.
    "A Fine homebuilding book"—T.p. verso.
    Includes index.
    ISBN 1-56158-135-6
    1. Counter tops. 2. Laminated plastics. I. Title.
TT197.5.C68K56   1996   96-9014
684.1'6—dc20   CIP

This book is dedicated to my early mentor, Bruce Womer, his sweet and gracious wife Patty, Robert the mason, Bear the grizzly carpenter, Harvey the ex-biker homesteader, and that long ago Vermont summer of '77, when I first realized I was born to build.

ACKNOWLEDGMENTS

This book has come to fruition because of the contributions of many people. My sincere thanks go out to the following:

Clancy Edmonds, a talented craftsman and great teacher, who over 15 years ago introduced me to the basics of working plastic laminate;

Jack Wellauer, who gave me the opportunity to make many countertops and put up with my shortcomings as I learned the craft at his expense;

Kevin Ireton, editor of *Fine Homebuilding* magazine, who encouraged my early writing attempts;

Jefferson Kolle, another fine *Fine Homebuilding* editor, who took an interest in my idea to write a book on plastic laminate and interceded on my behalf with the books editor;

Julie Trelstad, Fine Homebuilding books editor, who liked my idea, heralded it to the "higher-ups" and guided the project to completion;

Karen Liljedahl, editorial assistant, who assisted with many behind-the-scenes details;

Ruth Dobsevage, who so aptly edited my rough manuscript;

Steven Bossard, my good friend and work partner, who graciously contributed so much of his time and effort helping to get the photographs;

John Michaels and David Hall, who reviewed my manuscript and offered their encouragement and advice;

Marlene Kimball, my loving wife, who allowed me the space and time I needed, and patiently endured my "mental absence" from the family, as I wrote the manuscript.

My thanks also to all the industry and product representatives I consulted with, as well as the many other people at The Taunton Press who had a hand in putting it all together.

CONTENTS

INTRODUCTION

As I write this, the U.S. domestic market is consuming 1.4 billion sq. ft. of plastic laminate a year. While not all of that material goes into countertops, enough of it does to make plastic laminate the single most popular countertop surfacing material in this country, and for that matter, the world. It is a position laminate has held for decades, and, despite the challenge from Corian-type solid-surface countertops, there is no doubt that laminate will retain its top place (literally and statistically) for a very long time. The reason for this is simple. Plastic laminate is relatively inexpensive, attractive, durable and versatile. There really is no other material quite like it.

I was introduced to the basics of working plastic laminate as a carpenter's helper when I was a young man starting out in the building trades. Since then I've fabricated hundreds of laminate assemblies, most of them countertops. Over the years I've come to love working with this unusual product. However, from the very start I realized that plastic laminate was an exacting and sometimes unforgiving material. Mistakes are easy to make and seldom easy to correct, especially for those not familiar with the nuances of the craft. Unfortunately there has always been a lack of good how-to information about working laminate, and in my early days I suffered my share of setbacks. But I took solace in the fact that every mistake was a lesson learned, and I have learned my lessons well.

With knowledge and experience comes confidence, and gaining confidence is what this book is all about. Read it, and you will learn about the tools and techniques I use to fabricate with laminate and gain the benefit of my experience. If you're a novice, this book will serve as a complete countertop fabrication course, encompassing the basics and much more. If you have some experience with laminate but lack expertise, this book will enhance your skills. Even if you're a seasoned plastic-laminate fabricator, you can still glean some valuable nuggets from these pages. Most professionals are eager to learn how other professionals work, and are willing to try new tools and techniques in an effort to become even better at their craft.

This book focuses on teaching you the craft of making plastic-laminate countertops. That, after all, is the material's best known and arguably its most functional application. However, as I'm sure you're already aware, plastic laminate is also used in the manufacture of cabinets, furniture and commercial store displays, as well as for many other decorative and useful purposes. The possibilities are truly unlimited. I even know of a bathroom that has a pink plastic-laminate ceiling. Although a pink laminate ceiling may not appeal to you (it actually looks quite nice), it's likely that someday you'll want to use laminate for something other than a countertop. If so, you'll be glad to know that this book will provide you with the basic information you need to tackle almost any laminate challenge that comes your way. That's because laminate is laminate, and regardless of what you use it for, the same principles, tools and techniques apply.

I have written this book from my perspective as a professional small-scale fabricator. I make all my countertops by myself or with one helper, either in my workshop (which is the size of a two-car garage) or, more often, in place at the customer's home. I don't have a lot of expensive specialized equipment because I don't need any to produce a top-quality product, and you don't either.

The only thing I can't do that the big shops can is make post-formed countertops. A post-formed top is made by forming one continuous piece of laminate over an integral backsplash, across the counter and over the front edge. Such countertops have distinctive curved edges where the laminate is bent, and are made with the use of expensive forming machines and presses. The custom square-edge tops I make have separate pieces of laminate applied to the backsplash, countertop and countertop edges. The backsplash is made separately and then attached. While post-formed tops have the advantage of no edge seams, they have several shortcomings which I'll touch on throughout this book. However, I will not be discussing the specifics of post-formed countertop production.

With plastic-laminate work, you should always prepare yourself for the inevitable fabrication faux pas; absolute perfection in this craft is elusive, especially at first. In fact, I once believed that a perfectly sized, seamed, fitted, trimmed and filed countertop was downright impossible to achieve. However, I've never been satisfied with mediocrity, and have since proven myself wrong; the flawless top is not impossible, just rare.

The important thing, though, is to avoid making serious errors, and if you follow the advice in this book, that shouldn't be a problem. As for the little glitches and finishing blemishes (like filing an edge too far), they're a definite bummer, but they're minor; with experience they will become less frequent. Until then, take your lumps, keep your chin up, and press on.

Because there is often more than one way to do a job right, I will also tell you about legitimate alternative fabrication techniques. But the methods I emphasize are those that work best for me. In time, as you pursue this craft, I'm sure you will develop your own opinions and style of work, and that's good.

This book assumes that you already have basic hand and machine woodworking skills. I won't tell you how to measure accurately or how to use chisels, block planes, clamps or squares. Nor will I explain how to operate a table saw, a portable circular saw, a router or a drill. If you lack these skills, you will need to develop them before attempting to do anything I'm about to show you.

To get the most value from this book, read it through completely first. Once you have an overview of the fabrication process, you can refer back to those sections that you need. Don't hesitate to underline pertinent sentences. Keep the book close by when you're making your first countertops; my feelings won't be hurt if you spill contact cement on the pages, or figure material lists in the margins.

The book begins with a survey of the three main materials you will need for your countertop: plastic laminate, underlayment and adhesive. Chapter 2 deals with designing your countertop and estimating materials. Chapters 3 and 4 will tell you how to build the underlayment and rough-cut the laminate. Chapter 5 covers gluing down the laminate and trimming the edges. Chapters 6 and 7 cover backsplashes and custom edge treatments; Chapter 8 will tell you how to resurface worn or outdated countertops. Additional information on specific tools and products can be found in Resources, at the back of the book.

Once you've read this book, you will have the technical background and practical how-to you need to fabricate your own plastic-laminate countertops, and I wish you the best of success in your fabrication endeavors.

If I can be of further assistance by answering any particular laminate fabrication questions you have, I'd be more than happy to do so. Please direct your queries to me in writing c/o Fine Homebuilding Books, The Taunton Press, P.O. Box 5506, Newtown, CT 06470-5506, and include a self-addressed and stamped envelope for the reply. I will get back to you as promptly as possible. I'd also be pleased to hear about any unusual fabrication experiences, solutions to problems, ideas for custom edges, fabrication tips and techniques, and comments on tools or products. Aside from my personal interest in these things, your input will be most welcome when it comes time to update this book.

A Word About Safety

When I was in high school I knew a kid who could put his index finger up his nostril all the way to the second joint. I was especially impressed the first time I saw this feat because I could see he wasn't trying to fake me out by bending his finger out of sight; it really was in there that far. When he finally pulled it out I realized his rare display of anatomical impoliteness was possible only because the exact amount of finger that appeared buried in his nose was, in fact, missing. He had chopped it off with a radial-arm saw a couple of years earlier.

Shop safety, however, is not a joke. Over the years, I've rushed numerous co-workers to the doctor for various injuries, and have had some job-related injuries of my own—fortunately, never serious enough to require emergency medical attention. In every instance the accident was due to a moment of inattention or a careless disregard for common-sense safety precautions. Woodworking and building are inherently dangerous no matter how experienced you are. When working with hand or power tools, stay focused on your work, keep the area well ventilated and follow standard safety procedures at all times.

1

MATERIALS

One thing I've learned during my years in the building trades is that every project, no matter how complex it may appear initially, is just a series of smaller, simpler projects. If broken down to its elements, even the most difficult job becomes first conceivable and then, when successfully completed in logical progression, achievable. With that thought in mind, we'll begin with a discussion of the three main elements of a plastic-laminate countertop: plastic laminate, underlayment and adhesive. After you familiarize yourself with this basic information, you can confidently move ahead to designing and planning your project, and then to the actual construction.

Plastic laminate used for fabricating countertops is a sheet material that lacks structural strength. Therefore it must be applied over a solid, supporting underlayment of wood, plywood or particleboard. An adhesive, usually contact cement, is used to bond the laminate to the underlayment. Now let's take a closer look at each.

PLASTIC LAMINATE

When Daniel J. O'Conor, a 31-year-old Westinghouse chemical engineer, invented plastic laminate back in 1913, he had no idea the material would one day be used for countertops. But he did know it was an excellent electrical insulator, and O'Conor believed it could prove useful to the burgeoning American electrical industry.

Westinghouse's response to the breakthrough amounted to a corporate yawn, so O'Conor and another engineer, Herbert A. Faber, in a classic example of bootstrap entrepreneurship, went into the business of manufacturing plastic laminate on their own. Their first order, from the Chalmers Motor Company, was for commutator V-rings for electric motors. Previously, mica had been used for this purpose. Because plastic laminate was a substitute for mica, the new company and the material it produced were dubbed "Formica."

Over the years, the quest for practical applications for plastic laminate led the Formica Company to produce such diverse items as automobile timing gears and aircraft control cable pulleys. It wasn't until 1927 that the first primitive sheets of decorative plastic laminate appeared. The rest, as they say, is history. In the postwar building boom of the late 1940s and early 1950s, Formica rode the crest of the wave, and plastic laminate became a mainstay of the building industry.

Although Formica was the company that invented and pioneered the uses of plastic laminate, it is now only one of many manufacturers of this material (see Resources on pp. 127-129). However, Formica has developed such incredible name recognition that in the minds of the general public, all plastic laminate is Formica. Or, to put it another way, Formica is Formica, the stuff they make countertops out of; what's plastic laminate?

Without a doubt, the Formica Corporation must love the universal recognition of its name, because it's great for business, but its legal department takes a dim view of using the Formica brand name as a general term for all plastic laminate. In this book I will stick to the term plastic laminate, or simply, laminate. You should be aware though, that plastic laminate may also be referred to as mica (an obvious derivative of Formica) or HPDL (high-pressure decorative laminate, the technical term used in the laminate industry).

How plastic laminate is made

While each manufacturer probably has its own special "recipe" for making plastic laminate, the process is basically the same for all brands. Several sheets of brown kraft paper (the same paper used in grocery bags) are saturated with phenolic resin (plastic) and stacked in a press. A sheet of decorative paper is layered on, and finally a top layer of melamine (another kind of plastic resin, and one of the toughest plastics available) is added to the pile. The plastic and paper sandwich is subjected to high pressure and heat for a period of time, which causes the resins to flow together and bond into a solid sheet of laminated plastic. When the process is complete, the sheets are trimmed to size and the backs are sanded.

Plastic laminate is made of several layers of brown kraft paper and topped with a decorative paper layer. When the components are impregnated with plastic resins and subjected to heat and pressure, they fuse into a solid sheet. The curled foil is used in the pressing operation and is not part of the final product (shown on top). *Photo by Susan Kahn.*

Vertical, post-forming and horizontal laminate grades

There are three general-purpose grades, or types, of plastic laminate: vertical, post-forming and horizontal. Vertical grade is the thinnest of the lot and is designed to be used for non-work-surface applications such as on cabinet doors. I have been told that there are fabricators who use vertical-grade laminate for countertops, but the material is really too thin for this application and not recommended.

Post-forming laminate is a bit thicker than vertical grade, and has a different resin formula that allows the sheet to be heated and bent. It will bend to conform to a tight radius (typically down to ¼ in.) when heat (typically around 350°F) and pressure (typically around 1,000 psi) are applied with the proper equipment (typically a few thousand dollars worth).

Post-forming laminate may be used for custom square-edge fabrications, but it does cost a bit more than the other general-purpose grades.

Horizontal-grade laminate is made specifically for custom square-edge countertops and is therefore the preferred type for the work described in this book. Horizontal-grade laminate has greater impact resistance than post-forming grade and usually comes in a greater variety of sizes. On a countertop with inside corners (an L- or U-shaped top), horizontal grade is a must because it will resists inside-corner stress cracking better than the other grades (more about this on p. 53). Because horizontal-grade laminate is thicker than vertical-grade laminate, minor irregularities in the underlayment are less likely to telegraph through to the finished countertop surface.

General-purpose laminate should not be exposed to temperatures exceeding 275°F. It should not be used in exterior applications. And it should not be applied directly over concrete, plaster or drywall (for more about laminate and drywall, see pp. 107-108).

Special-purpose laminates

The special-purpose laminates mentioned here are available from any laminate supplier, though they are seldom stock items. All but backer sheet are substantially more expensive than standard grades. I don't intend to discuss these laminates in detail, but only want to introduce you to their availability and uses.

Fire-rated decorative laminate is most often used in commercial situations where the fire code requires it. I've never had a need for it.

Chemical-resistant decorative laminate is designed for use in hospitals, laboratories and other places where exceptional resistance to chemicals (including acids) and stains is necessary. (While general-purpose laminates have considerable resistance to many chemical solutions, berry juices and fabric dyes will cause staining, and bleach and hydrogen peroxide may damage the surface.)

Solid-color decorative laminate does not have a dark-colored phenolic back; the surface color is homogenous. With this type of laminate, the typical dark brown phenolic backer stripe that shows up along countertop edges (particularly with light colors) is eliminated. Solid-color laminate is worked much the same as basic horizontal grade, but a few areas, such as gluing along the edges and surface seams, require special attention. I will not be discussing the particulars of working with solid-color laminate in this book because my experience with it is limited and I don't believe it's all that popular. If you intend to fabricate with solid-color laminate, get the technical details from the manufacturer.

Decorative metal laminates are available for non-work-surface applications; they are never used for countertops. Thin brass and aluminum layers are bonded to a plastic-laminate base, allowing the sheets to be machined and glued down with basic laminate fabrication techniques. Metallic gold, chrome, bronze and copper colors are available as well as mirror-quality finishes. Here again, special fabrication instructions should be obtained from the manufacturer before using metallic laminates.

Wood-veneer laminates are also available. While general-purpose decorative laminates can be had in a vast array of ersatz wood-grain patterns, it is also possible to buy real wood veneers bonded to a laminate backer. Such laminates would not be desirable for the typical countertop work surface but would work well for cabinet skins, furniture and paneling applications. The wood veneer comes natural and requires finishing.

Custom laminates are a recent innovation now offered by some manufacturers. Words, designs, symbols, and patterns can be made to a customer's specifications, and are manufactured as an integral part of the finished sheets of laminate. Custom laminates are made either by printing on the decorative paper or by a precise laser cut-and-fit of different patterns (laminate marquetry if you will) before the top layer of melamine is added and the sheets are pressed together.

Laminate backer sheet is nothing more than dark phenolic laminate without the decorative top and melamine surface layers. In a laminate assembly that will not be securely fastened down all around (i.e. a pedestal table in a restaurant) it is important to install a sheet of laminate on the bottom to seal the assembly and minimize the potential for warpage that can result from unequal dimensional movement between the top and bottom surfaces (see pp. 10-11). Backer sheet is sold specifically for this purpose. Some fabricators use backer sheet on all countertops, but this is not a common practice. Since most countertops are (or should be) fastened down to the cabinets or framework below, they are physically restrained from low-level dimensional movement, and the backer sheet is optional.

Colors, textures and brands

When I was a kid in the early 1970s, orange countertops were the rage—even the Brady Bunch had an orange countertop. My family didn't have that color counter, but my parents did the next worst thing—they painted the kitchen walls a bright gloss orange. With such a heritage, you can understand why I am reluctant to advise my customers to select a particular design or color for their countertop. However, I do offer a few cautionary guidelines for work-surface counters that simplify the selection process and will lead to a better-looking counter for many years.

Avoid orange and other bold colors like red and yellow. No matter how trendy they are when new, such colors are overpowering and will lose their appeal with the passage of time. Besides, bold colors can be difficult to color coordinate with other decorative items in the room.

Avoid dark, solid-color tops because a monochrome surface highlights every speck of dust and is therefore almost impossible to keep clean. Moreover, minor scratches in the melamine top layer will always show up white (deep scratches that reach into the dark phenolic backer will, of course, show up dark).

Avoid gloss surfaces. Although a gloss finish looks super on the sample chip (and, for that matter, on a brand-new countertop), it is extremely impractical for a work-surface counter. It shows scratches readily, it reflects under-cabinet lights like a mirror, and it will highlight the slightest of imperfections (like a small errant chip of sawdust inadvertently glued in under the laminate). Also stay away from rough patterned textures because their nooks and crannies hinder easy cleanup. A matte finish is easier to fabricate with, and it will look better for longer than any other finish.

Today, there are literally hundreds of colors and patterns of plastic laminate to choose from, and most styles are also available in a variety of surface textures. (One company, Wilsonart, currently offers 240 different color and style patterns, most available in any of seven different textures.) Textures range from a perfectly smooth high gloss to a minutely textured matte to a finely

Plastic laminate is available in a dizzying array of colors, patterns and textures. Sample chips like these are available from all manufacturers.

pebbled gloss texture called Touchstone. I've also seen rougher patterned textures like slate, leather and geometric grids.

I'm often asked what brand of laminate is the best. Over the years I've used five different brands of laminate and have concluded that they are for the most part comparable in quality; there are no clear-cut forerunners. However, I have my favorite, and it's the Wilsonart brand, mainly because I have a nearby supplier who stocks a large selection, delivers on a weekly basis to my shop, and sells it to me at a reasonable price.

If you intend to do a lot of laminate work, it's a good idea to seek out a dependable distributor and buy your materials wholesale. Otherwise, most lumberyards or home centers will have sample chips for you to choose from, and you can order whatever you want. Most standard colors and patterns should arrive in less than two weeks. An unusual pattern, finish, or special-purpose laminate may take longer (I've waited as long as six weeks) because

sometimes the manufacturer will have to make the laminate to fill your order. So plan ahead, check with the distributor about availability, and order your laminate in plenty of time, especially if you have a tight schedule or a deadline to meet. One nice thing about laminate sheets is that they can be rolled up, boxed and shipped via UPS, which means that if a nearby supplier doesn't have what you want but a warehouse outside your area does, you will be able to get it relatively quickly.

Whenever you buy a sheet of laminate, inspect it carefully for breakage, surface scratches and pattern imperfections. I've encountered all three, and if I can work around them, I will. Otherwise I request a replacement sheet. The important thing is to make this inspection before you take possession of the material, not when you're about to use it.

Storing plastic laminate

Laminate manufacturers recommend that their product be stored inside and horizontally flat, with the top sheet turned face down and covered with a protective caul board. You definitely want to store the material inside, and, if you have room, by all means follow the rest of the recommendation. But laminate sheets come in sizes as big as 5 ft. by 12 ft. (see the drawing on p. 26), and the material is relatively brittle, so when space is limited, you will need to devise a more practical storage method.

In my shop I have an area against an outside wall where I keep a small supply of particleboard and plywood stored vertically (with the 4-ft. width on the floor and the 8-ft. length upright). I store 4x8 or smaller pieces of laminate between the plywood pieces. Larger sheets of laminate are rolled into a cylinder and set out of the way wherever I can find space (see the sidebar on the facing page). Several sheets can be rolled separately and nested together.

Storing plastic laminate rolled up is a bit controversial. Some people worry that if it's stored that way for very long it will not unroll flat when you need to use it. This is a legitimate concern, but unless the sheet was rolled up real tight for a long period of time, I have never found it to be much of a problem; the little bit of curl will relax sufficiently if you unroll the sheet to let it acclimate before using it. I once used a 4x8 sheet of laminate that had been rolled up and stored in the rafters of a garage for 30 years (my customer got a good deal on it when a lumberyard went out of business). After all that time the sheet had developed a serious scoliotic "set," but it was still limber enough for me to fabricate a semi-circular take-out counter for a pizza parlor.

When possible, I have laminate rolled by my supplier, whose specially made table jig enables one person to roll the material up and tie it off with relative ease. Otherwise, if I have to roll a sheet, I first lay it across my workbench, good side up. Then, with a helper on one side of the sheet and myself on the other, we'll carefully bring one end up, tuck it down, and slowly roll. It's not an easy process because the cylinder is unwieldy and slides around, and the risk of breaking the sheet is ever-present. But I don't try to roll it tightly, and that makes it easier.

I've used everything from duct tape to nylon string to fasten together rolled-up sheets of laminate. Duct tape is great stuff, but it leaves an adhesive residue on the laminate that has to be cleaned off later. The string by itself is a bad idea because it will cut into the exposed end of the roll. Any thin strapping material will do the same, but if a protective piece of cardboard is first folded over the laminate edge, the pressure from the strapping is spread out and there is no problem. Apply at least two, and preferably three, straps to hold the roll together.

Humidity, temperature and grain direction

From a fabricator's point of view, one of the most troublesome characteristics of plastic laminate is its dimensional reaction to temperature and humidity. As we all learned in grade school, heat makes materials expand, and cold makes them contract. Plastic laminate is not exempt from this physical law. However, it is the humidity factor that a fabricator must be most wary of.

Humidity is molecular water vapor (moisture) in the air. Warm air holds more moisture than cold air, and air at a given temperature can hold only a certain amount of moisture. The term *relative humidity* refers to the amount of moisture in the air, expressed as a percentage of the maximum amount of moisture it can hold at that temperature. For example, 30% relative humidity means the air contains 30% of the moisture it's capable of holding.

Humidity can be a fascinating topic, involving not only deeper meteorological considerations, but also human health and comfort. As laminate fabricators, however, we are interested mostly in how humidity affects the dimensional integrity of wood and wood by-products. What does all this have to do with plastic laminate, you might ask. Surprisingly, quite a lot.

Despite its plastic-resin content, laminate, because of its high paper content, is primarily a wood by-product. And even though the wood has undergone an extensive manufacturing process to be turned into paper, its cell structure remains the same as the wood that it came from.

THE EFFECT OF HUMIDITY ON WOOD AND LAMINATE

When wood or plastic-laminate cells absorb water, they swell considerably more in width (cross grain) than in length.

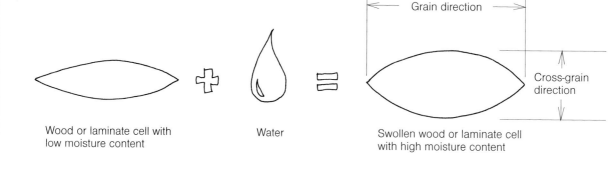

Wood or laminate cell with low moisture content

Water

Grain direction

Cross-grain direction

Swollen wood or laminate cell with high moisture content

GRAIN DIRECTION IN WOOD AND PLASTIC LAMINATE

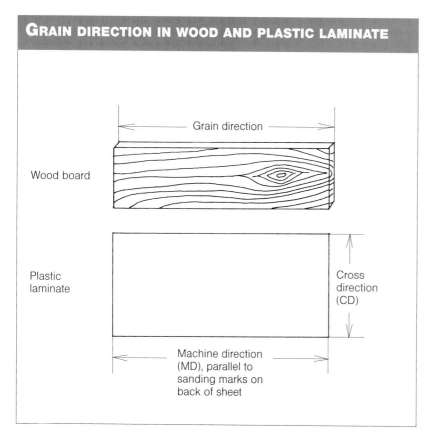

Wood board

Grain direction

Plastic laminate

Cross direction (CD)

Machine direction (MD), parallel to sanding marks on back of sheet

Like furniture makers, plastic-laminate fabricators must be concerned with two fundamental properties of wood cells. Wood cells are hygroscopic, which means they readily absorb and release water vapor in an effort to maintain equilibrium with the relative humidity of the surrounding air. And when the cells absorb water, they expand (see the drawing above). A wood board swells when the cells absorb moisture and shrinks when the cells lose moisture. Due to the shape and alignment of the cells, the greatest dimensional change is across the grain (usually in the width of the board). Although the hygroscopic process can be slowed by sealing the wood, it can never be stopped completely.

A sheet of plastic laminate is surprisingly similar to a wood board in that it has the same wood cells, and even a grain direction (see the drawing at left). The grain in a sheet of laminate is called the machine direction (MD), and you can tell the MD of a sheet of laminate by looking at the back of the laminate; the manufacturer's sanding marks run parallel to the grain of the paper, the long dimension of a full-size sheet. (Yes, paper has a grain direction. Try ripping a newspaper into strips and you'll find it works considerably better with the grain.) Cross direction (CD) is the direction perpendicular to the grain.

Grain direction is an important distinction because the National Electrical Manufacturers Association (NEMA) sets standards for dimensional stability of plastic laminates (see the sidebar at right). For horizontal-grade laminate, permissible limits are 0.5% maximum dimensional change in the MD of the sheet and 0.9% in the CD.

It's a great thing to have such guidelines, but what it all boils down to is that a 4-ft. by 8-ft. sheet of laminate can, depending on atmospheric conditions, still change dimension up to around ½ in. each way (larger sheets will move even more). The dimensional standard for post-forming laminate is even greater: NEMA allows 1.1% MD and 1.4% CD. Either way, we're looking at a whole lot of potential movement.

Coping with movement The trick to dealing with all of this undesirable movement is to recognize the very real possibility for disaster, and take the necessary precautions. Fortunately, like conscientious furniture makers who design their creations to allow for natural wood movement, we can do things to deal with the unstable nature of our material, and thus avoid the three most likely bad scenarios: surface seams that gap, laminate that shrinks back from the edges, or a completely destroyed glue bond.

LAMINATE PERFORMANCE STANDARDS

Although the useful applications for plastic laminate have long since transcended its humble beginnings as an electrical insulator, performance standards for the material are still set by the National Electrical Manufacturers Association (NEMA). My 1991 issue of the NEMA Standards Publication LD3.1, entitled "Performance, Application, Fabrication, And Installation Of High-Pressure Decorative Laminates," has a big title, but is only 18 pages long, and short on content. However, the booklet does provide some information that is not included in this book, and it is the only "official" (if not definitive) publication I'm aware of on the subject of working with plastic laminate. A copy costs $21, and you can order one directly from NEMA, 2101 L Street NW, Washington, DC 20037; (202) 457-8400; fax (202) 457-8411.

One important rule for combating dimensional movement failure is to make sure the laminate is bonded to an underlayment, or substrate, that has a similar dimensional stability. Once bonded, the two materials will move in unison. For more on underlayment selection, see pp. 12-13.

Another important rule is what I call the "rule of 48." The substrate, adhesive and plastic laminate should be stored in the area where they will be fabricated for a minimum of 48 hours before fabrication. There should be free air flow around the components during this acclimation period. Allowing all the materials to reach equilibrium with the temperature and humidity of the surrounding environment before putting them together substantially reduces the chance of failure due to dimensional movement.

In addition to the above two rules, NEMA indicates that the optimum atmospheric fabrication conditions are 45% to 55% relative humidity, at a temperature of 65°F to 80°F. It's been my experience that the temperature and humidity ideal is not as critical to adhere to as the rule of 48, although you certainly don't want to be fabricating in an obviously hot or cold environment.

UNDERLAYMENT

Plastic laminate is merely the attractive skin of an assembled countertop. The form and foundation of the structure is its underlayment. And, as is the case with all foundations, using the proper materials is critical to the success of the entire project.

To be compatible with the laminate, an underlayment must have similar dimensional properties. This requirement precludes the use of wood boards because they have a high degree of dimensional instability; in time, the normal seasonal movement of the wood will cause the glue bond to fail. The underlayment must also be smooth and flat because any unevenness might telegraph through as a perceptible ripple in the laminated top.

Manufactured wood-composite products are inherently more stable than the raw wood they come from, and are available in flat, smooth surfaced sheets. Three such products, shown in the photo below, are plywood, particleboard and medium-density fiberboard, usually referred to as MDF.

Plywood

The use of plywood as a substrate material goes back to the earliest days of laminate countertop fabrication, when it was used exclusively. These days, plywood remains an acceptable substrate but is seldom used. One of the reasons for this is that only better-quality plywoods with smooth surface veneers are appropriate for laminate work, and such material is expensive. The relatively cheaper CDX and underlayment grades of pine plywood, which are used for sheathing and floors in residential construction, are not recommended for gluing laminate to because they have an uneven surface texture.

Top to bottom: Medium-density particleboard, medium-density fiberboard (MDF) and plywood with a smooth surface veneer are the only materials that are appropriate for building a countertop underlayment.

Pine plywood with a smooth sanded finish (A grade) could be used, but it's expensive compared to other acceptable substrates. And A-grade pine plywoods often have voids in the plies below the surface veneer, which amount to hollow weak spots below the finished top. Besides that, pine plywoods are a pain to work with because the edges tend to splinter during milling operations. Most cabinet-grade hardwood plywoods, like birch or maple, are better to work with, but even more costly.

So even though it is acceptable, I recommend avoiding plywood altogether; price and workability aside, particleboard and MDF happen to be more suitable because their dimensional response to temperature and humidity is closer than plywood's to that of plastic laminate.

Particleboard and MDF

Occasionally, a know-it-all customer will insist I use plywood because it's better than "that chipboard stuff" (this said with a scowl). I never argue with customers about such things but I can tell you with distilled professional certainty that particleboard and MDF make a more uniformly flat, smooth and dense substrate than plywood. They are less expensive as well.

Now, if you were to ask me if I thought particleboard would make a good floor underlayment in a bathroom next to the tub, or if it is best for the bottom of a sink cabinet where the plumbing may eventually leak, I would say no, because particleboard has notoriously little resistance to water. It will soak up water like a sponge, swell and rot away more readily than any other wood product I can think of. But, because the underlayment of a countertop gets covered with a layer of waterproof plastic laminate, there is rarely a problem when the countertop is constructed properly.

Particleboard is made of sawdust and fine chips of wood combined with urea glue. When heat and high pressure are applied to the mixture, a dense, homogenous and heavy wood sheet is formed. The product is known variously as particleboard, chipboard, flakeboard and coreboard.

Before I understood there were different grades of particleboard, I mistakenly used a sheet of low-density floor underlayment (the wood chips are bigger, the sheets are not as dense, and the surface is not as smooth) to make a countertop. That was a long time ago and it has held up perfectly well, but since then I've always used the more appropriate medium-density grade, which has a density range of between 40 and 50 psi (pounds per square inch).

The manufacturing processes for MDF and particleboard are similar, but with MDF the raw wood material is further broken down into individual fibers before being "reassembled" into sheet form. The result is a wood composition panel that's finer grained (therefore much smoother) and denser (even heavier) than particleboard.

MDF costs more than particleboard (though less than better grades of plywood) and is not readily available in many retail markets, so particleboard may be your best bet for a countertop underlayment. Most professional fabricators, myself included, use particleboard almost exclusively.

CONTACT ADHESIVES

Contact cement has always been the laminate fabricator's glue of choice. Its adhesive ingredients are suspended in a liquid solvent, which makes it easy to spray, roll or otherwise spread out a thin, even coating. The glue is applied both to

Contact cement is by no means the only (or even the best) adhesive for gluing plastic laminate, but, because it doesn't require clamping, it is the most convenient, and for most small-scale fabricators, it is the only adhesive even considered. However, it's worth noting that other adhesives will also serve the purpose.

Many industrial fabricators, with the necessary big presses (i.e., desk-top manufacturers), use polyvinyl acetate (PVA) glue, which is similar to white Elmer's. PVA makes for a stronger bond, but it lacks the water resistance of contact cement. I have experimented with using waterproof yellow wood glue (yellow wood glues are a modified PVA adhesive called aliphatic resin) on small clampable projects like backsplashes and shelving, and it works very well.

Other adhesives, like resorcinol, urea formaldehyde, epoxy, polyurethane and even cyanoacrylate (Super Glue) will work with plastic laminate too, and somewhere along the line you may want to try using them to some degree on a laminate project. But for most purposes, properly applied contact cement is entirely adequate.

the underside of the laminate and on the underlayment. When the solvent has evaporated sufficiently, the laminate is applied to the underlayment, and the two mating surfaces stick on contact.

Types of contact adhesive

There are many brands and kinds of contact cement (see the photo on the facing page), but they fall into one of three categories: flammable solvent, non-flammable solvent, and water solvent. Contact adhesives suspended in a water solvent are commonly referred to as being water based, while the others, containing solvents like toluene, hexane, ether and naphtha, are called solvent based. For information on particular products, see Resources on pp. 127-129.

Flammable solvent-based adhesives

Flammable solvent-based adhesives are the traditional standby; they've been around a long time. Their main advantage is that they dry quickly, but the solvent vapors are terribly noxious and, as the name says, flammable. Some adhesives are ominously rated on their containers as "extremely flammable."

With flammable adhesives, the freshly applied liquid solvent rapidly evaporates into vapors that can cause a violent flash fire if ignited. Appliance pilot lights, a burning woodstove or a lit cigarette in the area are all it takes to set off an explosion. If you use flammable solvent-based adhesives, make sure to extinguish all sources of sparks or fire in the area, and provide good cross ventilation. Furthermore, be aware that most flammable solvents are heavier than air; concentrations of adhesive vapors can settle down into a basement, where they remain volatile. And just as important, keep in mind that smaller beings, like children and pets, will, if allowed in the glue-up area, be exposed to much higher concentrations of vapors than adults in the same area.

Non-flammable solvent-based adhesives

Non-flammable solvent-based adhesives were the industry's answer to a market demand for safer glue. They retain the desired fast-drying properties of the flammable adhesives, but their vapors are not combustible. However, the vapors are still noxious and unhealthy to breathe (see the sidebar on pp. 18-19).

It's worth noting that use of the popular non-flammable adhesive solvent 1,1,1-trichloroethane (methyl chloroform) was suspended by federal regulations at the end of 1995. The chemical had been used for many years in virtually all non-flammable contact cements but was removed because of environmental concerns. As a result, some adhesive manufacturers have discontinued all non-flammable solvent-based adhesive production. Others have changed their "blend" by substituting another non-flammable solvent, methylene chloride.

There are a lot of adhesive types and brands to choose from. A water-based type is the healthiest choice because toxic fumes are minimal.

Water-based adhesives Because of growing health and environmental concerns, water-based adhesives have become more popular with laminate fabricators in recent years. However, there continues to be some resistance to water-based adhesives among veteran fabricators. Their hesitancy is, I suspect, a natural reluctance to embrace new products and building practices; as is often the case, they prove not to be the panacea they were originally thought to be. But water-based adhesives are not all that new. According to the people at 3M, that company's Fastbond 30NF "water-dispersed" adhesive has been around for three decades.

In my own work I now use water-based contact cements almost exclusively, and I've found that they work as well as solvent-based adhesives. Their only drawback is that they take longer to dry.

Factors like air temperature, humidity and cross ventilation affect the drying time of any adhesive, but especially the water-based ones. Under ideal conditions (low humidity, temperature around 65°F and good ventilation), water-based adhesives will dry in about 15 minutes. A half-hour is the average drying time. On occasion, I've waited more than an hour. Solvent-based adhesives will dry in half that time or less; some solvent-based adhesives dry in less than a minute.

While drying time may be an important factor to some fabricators (especially production shops), I don't consider the longer wait for water-based adhesives to be a problem. Sure, it would be nice if the stuff dried faster, but I've always found there are plenty of other productive things I can do while waiting.

There is one definite, albeit slight, disadvantage to water-based adhesives, especially in a cold climate. If subjected to freezing temperatures, they will be rendered useless. You'll know right away if your glue has frozen because it will solidify into a useless rubbery chunk. Solvent-based adhesives are usually unaffected by freezing temperatures.

There aren't a lot of water-based adhesives on the market, so choosing a brand is also a lot easier than deciding on a solvent-based type. One adhesive company, Lokweld, currently offers 12 different flammable solvent-based adhesives, ranging from various kinds of spray grade to brush and roller grade and post-forming grade, but only one water-based product. This is also the case among other manufacturers. I have used Lokweld's H_2O adhesive and 3M's Fastbond 30NF adhesive, and I recommend both. Either product is suitable for spray, brush, or roller application (and can be used for post-forming operations too).

Applying contact adhesive

Contact cement may be applied by spray gun, brush and roller, or aerosol can. Most large shops, and a few smaller ones, use a spray-grade adhesive that has been formulated to be applied with a spray gun. Traditional compressor-operated spray equipment is typically used, but the newer HVLP (high-volume, low-pressure) spray outfits can also be employed. Two big advantages to spraying adhesive are that it's fast and it allows the fabricator to apply an even, smooth coat, which is the ideal. So far I've avoided using spray equipment because I haven't been able to justify the capital investment. Besides that, since most of my tops are site built, it's not practical for me to haul the equipment from job to job, let alone deal with messy overspray in someone's kitchen.

Brush-and-roller grade adhesive is applied manually with—you guessed it!—a brush or roller. It may not be as speedy as spraying, but the brush and roller approach (the brush for edges and the roller for large open surfaces) is fast enough for the needs of the average small-scale, hands-on fabricator. And when applied properly, the layer of glue will be more than smooth enough for a good bond. For a detailed discussion of the application process, see pp. 73-76.

Some manufacturers offer contact adhesives in an aerosol can (see the photo on the facing page). It costs more than brush-and-roller grade cement packaged in paint-type cans, but it offers the advantages of spray equipment without a major investment. And it provides the added benefit of not having to clean up the equipment—you just toss out the can when it's empty.

A few years ago, in an article I wrote for *Fine Homebuilding* magazine *(FHB #75, pp. 60-65)*, I heartily recommended the aerosol approach, but the particular brand I used (Lokweld #1055) was pulled from the market a short while after the article came out. (It was a non-flammable solvent-based product, and the manufacturer was concerned about the deleterious effects of the solvent.) I was disappointed because the product was convenient, and the vapors weren't as oppressive as those of the brush-grade solvent adhesive I had been using. But the loss of #1055 prompted me to try a water-based product, and that proved to be a good move.

A water-based adhesive in an aerosol container could be a great combination, and 3M is currently testing such a product in the California market. Until that proves itself and becomes universally available, there are other solvent-based aerosols to choose from, and I use them often. Lest I sound hypocritical, let me explain. Although I'm totally sold on

Aerosol contact adhesives are convenient to use and offer the benefits of spray-on application without the expense of a spray-gun system. *Photo by Kevin Ireton.*

water-based adhesives, I'm not a purist, I'm a pragmatist, and pushing down on a nozzle is more convenient than using a brush for a few small pieces or a quick little project. Besides, there are a few occasions when using a solvent-based adhesive is absolutely a must, such as gluing laminate to laminate when resurfacing countertops (see Chapter 8).

If you decide to use an aerosol adhesive, make sure you get one suited for laminate work: it should say so on the can. Many lightweight contact adhesives on the market lack the heavy-duty bond strength you'll need for the job. I'm partial to 3M's High Strength Adhesive 90. Lokweld recently introduced its #800 aerosol, but I have yet to use it. Both of these adhesives are flammable solvent based.

Whatever adhesive you use, keep in mind that all contact cement has a shelf life, and it's usually about one year. Some containers have a clearly marked expiration date on them, others will have a coded date. If you have any doubts about the viability of an encoded adhesive, call the manufacturer for a code translation. In any event, it's not a bad idea to glue up a test block to make doubly sure you're getting an adhesive that will work.

VOLATILE ORGANIC COMPOUNDS: A THREAT TO YOUR HEALTH

When I was a kid in junior high school in the early 1970s, we had drug-education classes where, among other things, the hazards of sniffing glue were clearly spelled out. Until then I had never heard of such a thing and thought it the pinnacle of stupidity. But only a few years later I found myself inhaling massive amounts of fumes from noxious non-flammable solvent-based contact adhesive while fabricating plastic-laminate countertops. While the cans of adhesive recommended plenty of ventilation in the work area and clearly warned that breathing the vapors was dangerous, many of my countertops were built in small kitchens, or it was winter, and adequate ventilation meant too much cold air. No one I worked with took the warnings very seriously, and neither did I.

As a result, getting high on glue fumes became a regular part of the job. My co-workers and I would joke about the side benefits of our work as we'd crack the lid on a fresh gallon of contact cement and start to spread it on. However, as more and more glue was put down, and more solvent evaporated into the air, the initial intoxication would quickly give way to dizziness, headache and nausea.

Our problem, aside from ignorance and the illusion of youthful invincibility, was volatile organic compounds (VOCs). VOCs are part of virtually all adhesive solvents, and when the solvents evaporate, VOCs are released into the air in the form of molecular vapor particles.

When you breathe air laden with VOCs, the toxic compounds find their way into your bloodstream via the lungs. While your central nervous system suffers the immediate, obvious and usually temporary ill effects of occasional VOC overload, continued frequent exposure to VOCs will likely lead to debilitating (or worse) health problems down the road. All the health ramifications from cumulative exposure to toxic VOCs aren't known for certain, but it's safe to say that permanent damage to the nervous system, kidney and liver can be expected.

Knowing this, it makes darn good sense to protect yourself from VOCs, and there are several ways to do that. The first, and probably the most effective, measure you can take is to avoid using adhesives that are high in VOCs. VOC content is measured in grams per liter (gpl), and is usually listed on the adhesive container. If you don't find it there, ask your supplier for a MSDS (Material Safety Data Sheet) form for the product

or products you are considering (see "To learn more" on the facing page).

Most solvent-based adhesives have a VOC content around 550 gpl, while water-based adhesives are typically less than 20 gpl. Comparatively speaking, if inhaling fumes from solvent-based adhesive were the equivalent of getting hit in the head by a cannonball, inhaling fumes from water-based adhesive would be like getting hit by a BB. In other words, water based is still slightly toxic, but the vapors won't knock you over. That's the main reason I now use water-based adhesives.

The next prudent safety measure is an appropriate respirator mask, especially if you're using solvent-based adhesives. Ordinary dust masks with rubber-band straps are absolutely useless when it comes to stopping VOCs; remember, we're talking about molecular particles here. Only better-quality face masks with replaceable organic-vapor cartridges are effective at filtering out VOCs. These respirators are available at auto-supply stores, and many lumberyards and home centers also carry them.

Organic-vapor cartridges are filled with activated charcoal, which, figuratively speaking, grabs onto the passing VOCs and holds them tight. Eventually, the cartridges will lose their effectiveness, but you'll know they need to be replaced when you begin to smell the solvent through the mask. By using a particulate prefilter, you can entrap dust particles (another health hazard) before they reach the charcoal, and that will substantially extend the lifespan of the cartridges.

Even a good respirator, though, doesn't offer blanket protection from all VOCs and solvent hazards. For example, two VOCs, methylene chloride (now used in most non-flammable solvent contact adhesives) and methyl alcohol, will pass freely through organic-vapor cartridges. Also, many solvents can be absorbed through the skin, so you may want to use solvent-proof protective gloves—the same kind sold for use with furniture strippers. And it's just as important to protect your eyes with safety glasses when using solvents as it is when you are performing woodworking operations.

TO LEARN MORE
A Material Safety Data Sheet tells you the chemical composition and physical attributes of a product. To learn about the possible risks of using a particular solvent or solvent-based product, it behooves you to obtain that product's MSDS form. All suppliers are required by law to have these forms and provide you with a copy upon request. MSDS forms are a valuable source of information on health hazards, flammability and safe handling.

Once you've armed yourself with the MSDS form, you can get further valuable information about how to protect yourself from the various VOCs, as well as other job-site health hazards, by calling the National Institute for Occupational Safety and Health (NIOSH) at (800) 356-4674 and requesting a publications list. NIOSH is a government (tax-dollar supported) organization, and many of its materials are free of charge.

A cartridge respirator with activated charcoal filters will filter out most VOCs in contact cements. With particulate prefilters (not visible in this photo), the cartridges will last longer. Don't rely on a common dust mask, which provides absolutely no protection from VOCs.

DESIGN
AND PLANNING

Before you can build a countertop you must design and dimension it. This stage of the work entails working out the shape of the underlayment and the build-up strips and cleats that will lend extra support at the edges and seams, adding in the overhangs and allowing for ranges and sinks. You also must plan for the backsplashes and decide where the seams (if any) in the laminate will go. When all that is done, you will be able to calculate the materials you will need for the job and place your order.

EVOLUTION
OF COUNTERTOP
DESIGN

Laminate countertops have changed quite a bit over the years (see the drawing on the facing page). Early plastic-laminate countertops consisted of a ¾-in. layer of plywood underlayment that was first fastened down to the top of the cabinets, then topped off with a layer of laminate. A chrome strip nailed over the edges completed the assembly.

In time, the chrome gave way to plastic laminate (called a self edge). The ¾-in. single layer of plywood remained, but the illusion of a thicker edge (about 1½ in.) was achieved by nailing a wide pine banding strip to the underlayment edge, with the excess hanging down. In an alternative design, a small batting strip added under the underlayment

achieved the same effect. These drop edges, as they were called, were suited to building countertops in place on traditional-style face-frame cabinets that were also built in. Although some custom fabricators today remain faithful to the style, drop edges are no longer a common construction method.

THE MODERN
BUILT-UP
COUNTERTOP

A typical countertop today still has ¾-in. underlayment (though it's usually particleboard, not plywood) as well as a 1½-in. thick edge. But the edge no longer "drops" over the cabinet face frame. Instead, wide ¾-in. thick strips of wood called build-up strips (sometimes they are called build-down strips) are attached to the bottom on all edges of the top sheet, which is called the deck, or decking. Where there is a joint between two pieces of decking, an extra-wide build-up strip called a cleat is fastened underneath to reinforce the seam. (For more on cleats, see pp. 46-47.) Intermediate build-up strips are also sometimes positioned between the front and back build-ups; they rest on the top of the cabinet sides and provide extra support. The top drawing on p. 22 shows a typical modern laminate countertop.

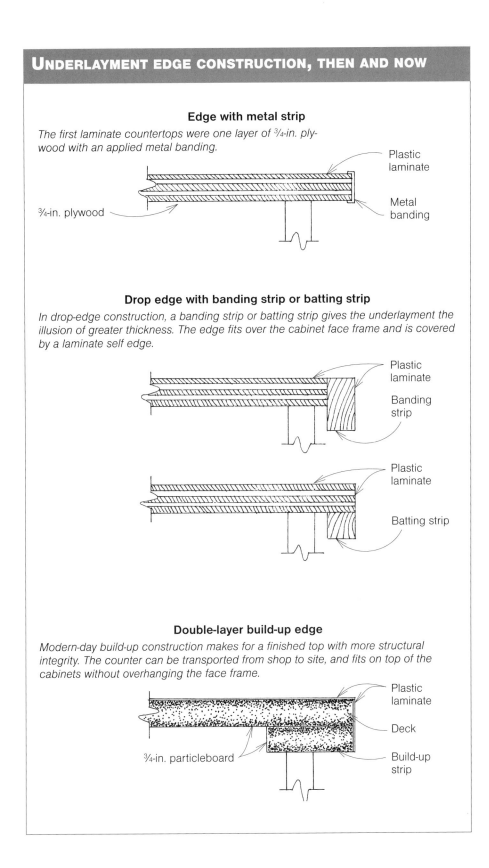

Edge with metal strip

The first laminate countertops were one layer of ³/₄-in. plywood with an applied metal banding.

Plastic laminate

³/₄-in. plywood

Metal banding

Drop edge with banding strip or batting strip

In drop-edge construction, a banding strip or batting strip gives the underlayment the illusion of greater thickness. The edge fits over the cabinet face frame and is covered by a laminate self edge.

Plastic laminate

Banding strip

Plastic laminate

Batting strip

Double-layer build-up edge

Modern-day build-up construction makes for a finished top with more structural integrity. The counter can be transported from shop to site, and fits on top of the cabinets without overhanging the face frame.

Plastic laminate

Deck

Build-up strip

³/₄-in. particleboard

A MODERN COUNTERTOP UNDERLAYMENT

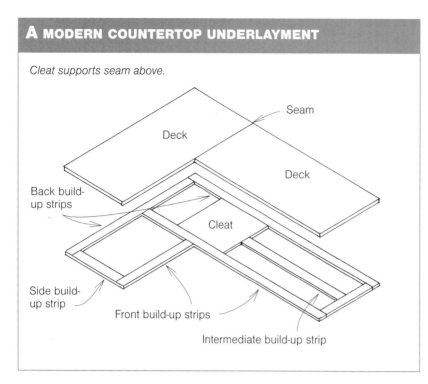

Cleat supports seam above.

Seam

Deck

Deck

Back build-up strips

Cleat

Side build-up strip

Front build-up strips

Intermediate build-up strip

BUILD-UP STRIP OPTIONS

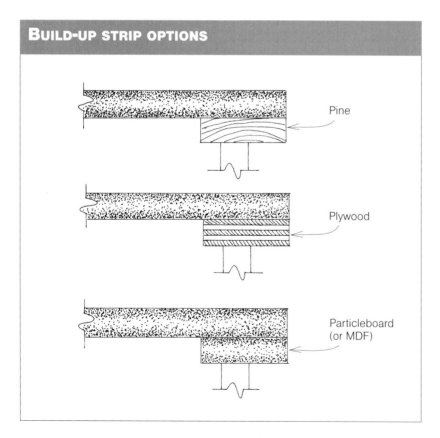

Pine

Plywood

Particleboard (or MDF)

Cleated double-layer construction offers certain advantages over earlier countertop styles. A built-up countertop underlayment is strong enough to be transported from an off-site production facility and then set in place. Also, a built-up countertop will not interfere with modern frameless cabinetry having full-overlay doors and drawer fronts, as a drop-edge counter would.

Build-up strips

Particleboard (or MDF) may be used for build-up strips, but strips of pine or plywood are legitimate options as well (see the drawing below left). The dimensional instability of a pine board is not a problem when only the ¾-in. thick edge of the board is being glued. A pine build-up gives the countertop more structural strength than a particleboard build-up. Also, it's not unusual near a kitchen sink for some water to escape over the edge of the counter and wick up into the build-up. Consequently, a particleboard build-up will swell a bit (though I've never seen it result in a serious problem), but the pine will take the small amount of moisture without any adverse effect. Despite these advantages, I seldom use a pine board build-up because it is more expensive than the other options, it has knots that sometimes create a problem on the edge, and it's more likely to split when fastened on.

Plywood makes an excellent build-up. I use ¾-in. underlayment grade because it's reasonably priced and has no core voids. Plywood has all the advantages of pine boards, but none of the disadvantages. I especially like to use plywood as a build-up when I have to make a cantilevered countertop that will provide at-counter seating, as on kitchen peninsulas or islands. In these instances I make the build-up big enough to fit under the entire overhang and extend back over the cabinets 6 in. to 12 in. This approach results in a solid overhang, as opposed to cantilevers on

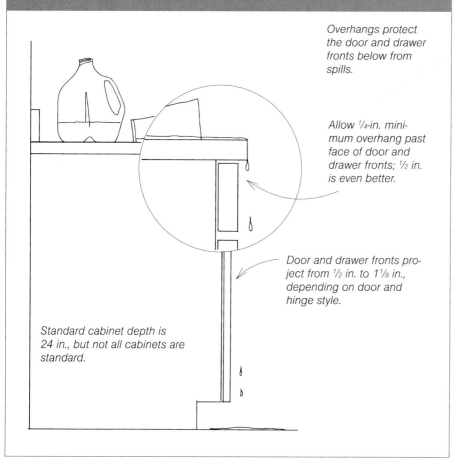

Overhangs protect the door and drawer fronts below from spills.

Allow ¼-in. minimum overhang past face of door and drawer fronts; ½ in. is even better.

Door and drawer fronts project from ½ in. to 1⅛ in., depending on door and hinge style.

Standard cabinet depth is 24 in., but not all cabinets are standard.

post-form tops, which are notoriously flimsy. I use a plywood build-up on large shop-built tops that I make because it has more inherent material strength than particleboard, and the whole assembly will be stronger, which is what I want if I have to transport a top.

On the other hand, a particleboard build-up is what most fabricators typically use, and it is perfectly acceptable. Most of my site-built tops are made with a particleboard build-up.

Sizing the overhangs

The average kitchen countertop over 24-in. deep cabinets is about 25 in. deep. The 1-in. front overhang shields the cabinet doors below from food crumbs and liquid spills that make their way over the edge (see the drawing above). However, I've known cabinet doors to project from the face of a cabinet up to 1⅛ in., so I always take into consideration the thickness of door and drawer fronts, then size the overhang at least ¼ in. wider; ½ in. is better. (Post-formed countertops come in one standard depth, and I have seen many post-form installations without sufficient overhang.)

The overhang often has to vary along a run of cabinets because many lengths of cabinetry are not perfectly straight, nor are corners perfectly square, and you certainly don't want the edge of a countertop to match a distorted run of cabinets. More overhang is preferable

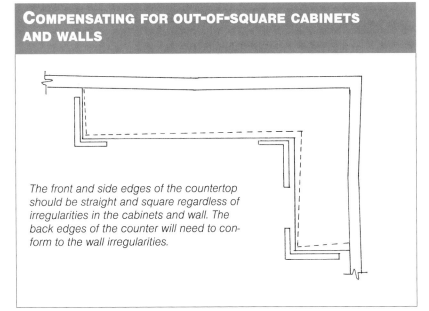

COMPENSATING FOR OUT-OF-SQUARE CABINETS AND WALLS

The front and side edges of the countertop should be straight and square regardless of irregularities in the cabinets and wall. The back edges of the counter will need to conform to the wall irregularities.

square inside corners; for details on how to make the fit, see pp. 60-65. Wall irregularities are a common problem in remodeling old houses, but they're not unheard of in new construction either.

In any event, few installations are as perfect as the illustrations in a typical how-to book would lead you to believe; variables are to be expected. But deviations are not a problem if you are making custom countertops; you can easily adjust the size and configuration of the tops to compensate for little (and sometimes big) inconsistencies. That is not the case with post-formed tops, which have very little fudge factor. More than once I've tried to fit a post-formed top on an L-shaped assembly of cabinets where everything was so out of whack that the overhang ended up being inconsistent and at some point insufficient, and the fit at the back wall was horrific. My disdain for post-formed tops stems in part from such tribulations.

Planning around ranges and refrigerators

Ranges require special consideration when planning and fabricating countertops. Drop-in units, which have no raised back on them and fit into a cut-out in the counter, are not usually a problem, as long as you remember that the laminate won't end at the sides of the stove; it will continue around behind, and you'll need to take that into account when figuring materials. Follow the range manufacturer's dimensions and directions when laying out the cut-out.

to not enough overhang to provide adequate protection for the cabinet fronts. The main objective is to have a straight and square-edged countertop (see the drawing above), and that will effectively obscure any inconsistency in the line of cabinetry below.

On end overhangs I aim for ½ in. to 1 in. past the cabinet side, unless there is some sort of restriction that demands less. If there happens to be a door or applied panel on the end or a run of cabinets, I maintain the same overhang as on the front edge.

On peninsula or island countertops, I usually maintain a uniform overhang on all edges. If there is to be an extended overhang to accommodate at-counter seating, 12 in. to 14 in. is considered to be the best width. And for comfortable seating, allow 2 ft. of counter length for every chair or stool.

Of course, the back of the countertop, against the wall, may need to deviate from straight and square in order to conform to imperfect walls and out-of-

A freestanding range with a raised back fits up against the wall and typically has countertop sections that abut it on each side. A precise stove-to-countertop fit is one of the hallmarks of a well-crafted custom countertop. The usual freestanding range is 30 in. wide, but I've seen "30-in." units that measured up to ¼ in. less. For this reason I prefer to have the range on site when I install the counter-

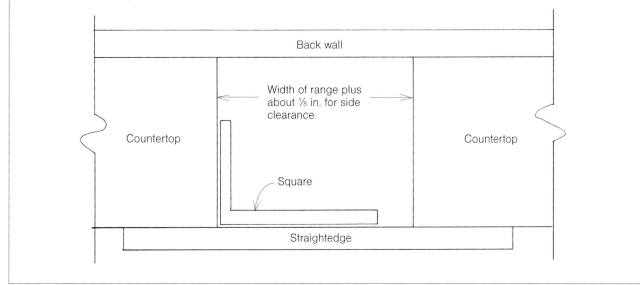

FITTING A COUNTERTOP TO A FREESTANDING RANGE

To get the best fit at the sides of the range, square the countertop sides off the front edges of the counter and make them parallel at the proper width. Don't try to square off the back wall.

Back wall

Width of range plus about ⅛ in. for side clearance

Countertop

Countertop

Square

Straightedge

top. I want the fit to be almost glove tight (¹⁄₁₆ in. space on each side), with the side spaces perfectly parallel with the edges of the appliance. The best way to achieve such tolerances is to slide the stove in place and fit the countertop to it.

If there is no range to fit to, get a reliable measurement, add the side clearance, and square the two counter sections off a straightedge placed up against the front edges of the countertops (see the drawing above). Never square your sections by placing your framing square against the back wall, since even a slight irregularity there will throw off the fit.

Rarely are countertops placed on both sides of a freestanding refrigerator, but if they are, it is usually best (if possible) to give the countertop a slight side overhang and make sure it's square to the front edge; glove fits and even spaces are not necessary or desirable here.

Planning for sinks

Except as they affect the location of seams in the countertop (see pp. 27-29) sinks are not a major issue at the planning stage.

BACKSPLASH DESIGN CONSIDERATIONS

Backsplashes are a definite design and planning concern. I will focus here on the design and planning details for a typical ¾-in. thick by 4-in. high attached backsplash. Other backsplash options and the specifics for installing them are discussed in Chapter 6.

Nowhere is it written that attached backsplashes must be 4 in. high, but somewhere around 4 in. seems to look best. The actual measurement may vary, depending on circumstances. For example, when replacing old drop-edge countertops that had no attached backsplash, you often find that the existing back-wall electrical receptacles end up a

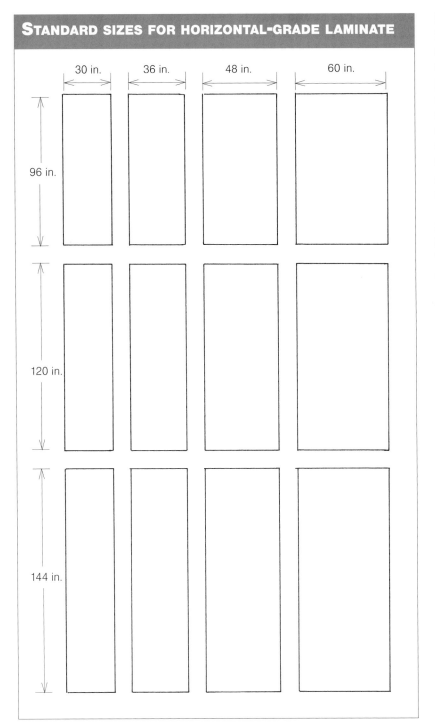

STANDARD SIZES FOR HORIZONTAL-GRADE LAMINATE

30 in. 36 in. 48 in. 60 in.

96 in.

120 in.

144 in.

bit closer than 4 in. to the new counter-top. If you rip the width of the back-splash down a little bit, the problem is solved. Conversely, if there are windows positioned close to a countertop, it's often possible to add an extra ½ in. or so to the usual 4-in. backsplash height and have the backsplash fit up tight to a window casing or the bottom of the window stool. This eliminates a narrow space that would look awkward and be difficult to paint or otherwise finish off.

SHEET SIZES AND SEAM PLACEMENT

With the major design details of the countertop worked out, you are almost ready to begin estimating materials. To figure the laminate needs for a particular countertop project, it's necessary first to know what sizes you can get. Armed with that information, you can then determine if any surface seams will be necessary. Ideally, you will want to elimi-nate seams entirely, but if this isn't possible, you'll need to plan their place-ment carefully.

Laminate sheets and self edge

General-purpose laminate is sold by the square foot and comes in a wide range of sizes. Available sizes will vary from manufacturer to manufacturer (and this is something you'll want to check into), but you can usually count on being able to get sheet widths of 30 in., 36 in., 48 in. and 60 in., and sheet lengths of 96 in., 120 in. and 144 in. (see the drawing at left). As a rule, sheet sizes will actually be ½ in. to 1 in. oversize in each direction. The little bit extra is a nice bonus and may come in handy, but I don't suggest you include it in any pre-project estimating plans.

It's also possible to buy 1¾-in. by 144-in. strips of laminate, which are commonly called self edge (see the photo on the facing page). If you're figuring your materials closely, or want to save the time and trouble of ripping narrow edge strips out of big sheets, self edge can be very useful. You will also want to buy self-edge strips if your countertop will have rounded corners with tight radiuses, because the strips are typically made of post-forming laminate, which is formulated to be heat-bent to tighter curves than horizontal-grade laminate (see p. 78).

Seams

The next step when figuring laminate material needs is to decide where to locate surface seams. With laminate sheet sizes available up to 5 ft. by 12 ft., any countertop configuration falling within that footprint can, and should be, fabricated without a seam. The ability to eliminate surface seams is one of the greatest advantages that custom countertops have over post-formed tops. Even if you end up with a lot of leftover laminate, I believe it's always better to use a big sheet if you can avoid a seam by doing so. There are many ways you can make use of the scraps (see the sidebar above right).

Most laminates have patterns and textures that are homogenous, so you can cut and seam anywhere you like and the pattern will match. However, sometimes there are very slight visual differences in the texture of the sheets when they are viewed with the grain (MD) as opposed to across the grain (CD). Often, this difference isn't readily apparent until two laminate pieces are seamed together with their grain directions at right angles. Then there will be a slight, though perceptible, variation in the tone, or shade, of color between the two.

Your laminate supplier should be able to tell you if the texture on laminate you're using is a directional one or not. If it is, the possibility of cross-grain tonal differ-

LAMINATE "RECYCLING"

Leftover pieces of laminate need not go to waste; often they can be put to use elsewhere. For example, excess material can be ripped into strips and used to laminate any window stools over the countertop; it's a nice design touch, and the laminate is more durable than wood. I do this by applying a batten-strip drop edge to the front projection of the stool to give it a thickness of about 1 1/4 in., then apply the laminate just as I would on a countertop.

Larger scraps can make bathroom vanity tops or be used to cover the plywood bottoms of sink base cabinets. I also use a lot of laminate scrap when making cabinet drawers. By gluing it to 1/4-in. lauan plywood (often left over also), I get cheap, durable and good-looking drawer bottoms.

Even small pieces can be put to good use. I cut up small pieces to use with glue test blocks or as disposable palettes when mixing seam filler (see p. 101).

These are just some ways I use the scrap laminate I end up with. I imagine you can think of several other applications.

Rolls of self edge are precut to width for edging laminate countertops. Self-edge laminate is post-forming grade and comes as 1 3/4-in. wide strips, 12 ft. long.

The fewer the seams in a countertop, the better the finished counter will be. But for post-formed countertops, seams are hard to avoid. Since post-formed countertops can be made only in straight sections, any counter configuration that changes direction requires two post-formed sections to be joined together with drawbolts on the underside. Making the connection can be a difficult job (especially with, say, a large U-shaped top fit against three walls) often made more frustrating when the pieces don't match up perfectly flush along the entire length of the joint.

Custom-made square-edge tops, on the other hand, can usually be made without fuss and compromise. The drawing below is of an actual seam-studded post-formed countertop that I replaced using one 5-ft. by 12-ft. sheet of laminate. The old post-formed counter had seven seams; the custom replacement, none.

Some laminates, like butcher block and other wood-grain patterns, have an obvious directional pattern that requires special attention. With such patterns, any seams placed in straight runs of the countertop should be made with the grain direction matching up. But when the countertop makes a turn, as on a U- or L-shaped countertop, the directional patterns will look good only if they meet in the corner at a diagonal seam. That means you'll have to make a seam where one might otherwise not have been necessary.

Directional wood grains are particularly difficult to work with because of the need for a diagonal seam, and also because the edge pieces need to be cut out and applied so that the grain pattern "flows" realistically over the edges. I've seen plenty of fake butcher-block laminate tops where these extra measures were not taken, and they look downright phony. I've even made a few like that myself in the past, and it's probably part of the reason I have developed such an aversion to fake wood in its many forms. Now I don't even let my customers see the wood-grain pattern sample chips, and hope they don't ask about them.

The diagonal corner seam is best for directional patterns but otherwise it's not necessary, unless you really want it. I try to avoid diagonal seams because they're long and obvious and, frankly, they're more difficult to make. Instead, I try to place seams where they're least noticeable (see the drawing on the facing page), and that's where there is a sink or range-top cut-out. In those locations, most seams are so inconspicuous that the countertop might just as well be considered seamless. However, I know of at least one fabricator who strongly disagrees with me on this issue; he believes that placing a seam in those areas (especially at sinks) is asking for problems

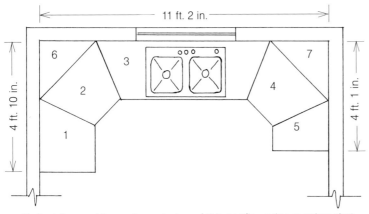

To install a post-formed countertop of this configuration requires that seven separate sections be joined together. Pieces 6 and 7 would be positioned above the main counter at the top of the backsplash (almost like corner shelves) and be supported in the corner with ledger strips underneath. This post-formed counter was replaced with a custom square-edge top cut from a 5x12 sheet of laminate. No surface seams were needed.

ences can be eliminated by matching the grain direction of sheets that meet at a seam. If that isn't convenient, position the seams through sink or range-top cut-outs where only an inch or two of seam is visible; any difference between the two sides will not be apparent.

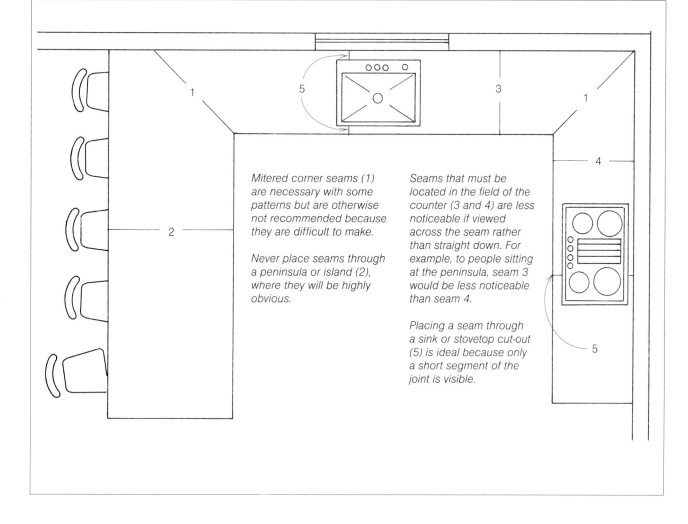

Mitered corner seams (1) are necessary with some patterns but are otherwise not recommended because they are difficult to make.

Never place seams through a peninsula or island (2), where they will be highly obvious.

Seams that must be located in the field of the counter (3 and 4) are less noticeable if viewed across the seam rather than straight down. For example, to people sitting at the peninsula, seam 3 would be less noticeable than seam 4.

Placing a seam through a sink or stovetop cut-out (5) is ideal because only a short segment of the joint is visible.

because, he says, the gap provides an avenue for water penetration into the underlayment. I understand his reasoning, but when I make a seam there is no gap (in Chapter 3 I'll show you how to do likewise), and I've never had a problem with seams by sinks.

When you can't get the seam into the span of a cut-out and it must be made across the depth of the counter, then I recommend you place it as inconspicuously as possible. One way to do this is

to keep in mind the predominant viewing angles in the kitchen. Most laminate seams (even, unfortunately, the well-made ones) are evident when viewed directly down the length of the seam. But when viewed across a seam (particularly from a distance), a seam visually disappears.

One last word on the placement of seams: make sure they fall at least 6 in. away from any seams in the underlayment decking.

ESTIMATING MATERIALS

Now that you have a handle on the various design guidelines, know the laminate sizes that are available, and understand the principles of seaming, you arrive at the nitty-gritty work of figuring out how much laminate, underlayment and adhesive to buy for your countertop.

Laminate

I wish I could give you a no-brainer estimating formula, but it's not that easy. Sometimes figuring the laminate materials is straightforward, but at other times it can be a challenge.

It's possible, of course, to take just a few basic measurements of a prospective countertop, order plenty of extra-big sheets, figure the details out when you do the job, and have lots of laminate left over at the end of the project. I have done that, but I've found any job will go much more smoothly (and cost considerably less) if I carefully figure the layout of laminate pieces before I start.

Here's my system for estimating laminate needs on a difficult-to-figure countertop I built last year for a new kitchen:

CABINET LAYOUT DRAWING

To figure materials needed for a countertop, start with a diagram of the cabinet layout.

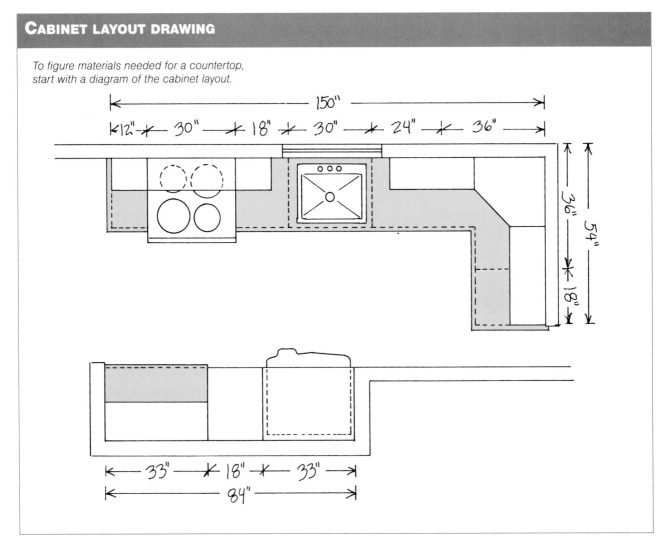

I start with a layout diagram of the cabinets (see the drawing on the facing page). For many tops (replacement tops in particular) I would need to take my own measurements on site and record them. Next, I draw the countertop to scale on ⅛-in. graph paper (see the drawing below). Each ⅛-in. square represents 2 in. I indicate the cabinet dimensions on the graph paper, and draw the countertop one full square larger (2 in.) on all overhanging edges. For example, if the cabinet depth is 24 in., I scale at 26 in. If the cabinetry measured an odd number, say, 25 in., I'd round up to the next even number (26)

and then add a 2-in. square. The figuring process is a lot simpler if you round up and keep all your measurements in 2-in. increments.

If I expect problems with wavy walls, or if the overhangs are oversize at some point, I'll scale on more squares. All countertop, backsplash, and other laminate components (such as the piece that will go on the wall behind the stove) are drawn to oversize scale on the grid—that is, all except countertop edge pieces and the backsplash top pieces; these components will be figured in later.

SHEET LAYOUT DIAGRAM

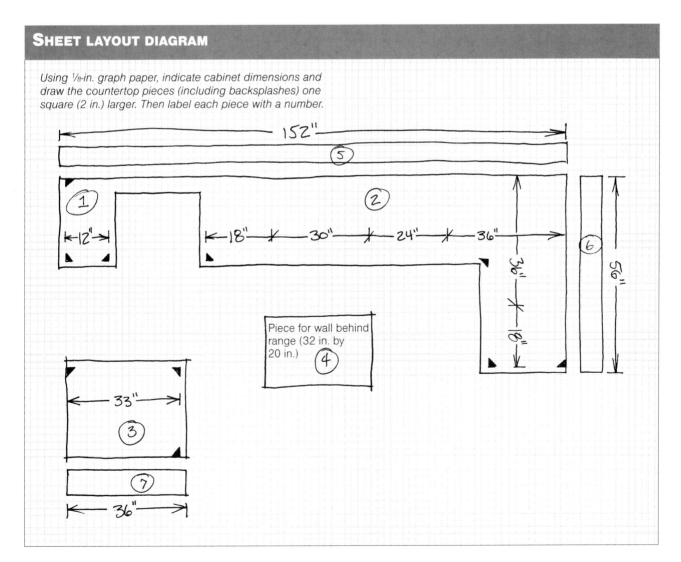

Using ⅛-in. graph paper, indicate cabinet dimensions and draw the countertop pieces (including backsplashes) one square (2 in.) larger. Then label each piece with a number.

152"

5

1

2

12"

18"   30"   24"   36"

36"

18"

56"

6

Piece for wall behind range (32 in. by 20 in.) 4

33"

3

7

36"

LAMINATE PLANNING SHEET

Photocopied laminate planning sheets, based on the same ⅛-in. grid as the scale drawing of the countertop, let you scale your pieces onto one or more 5x12 (or smaller) laminate sheets.

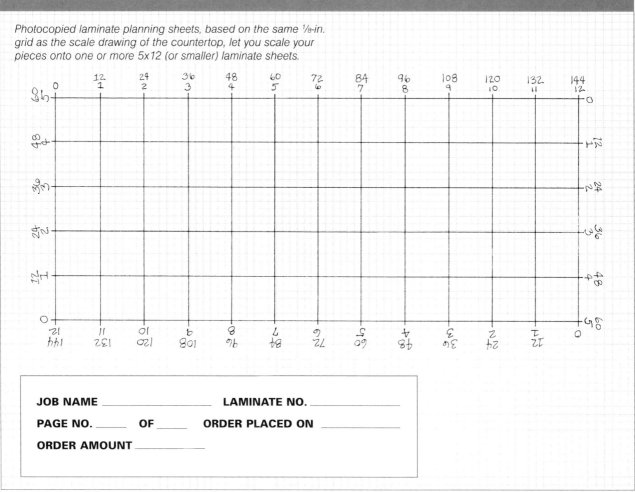

JOB NAME _____ LAMINATE NO. _____

PAGE NO. _____ OF _____ ORDER PLACED ON _____

ORDER AMOUNT _____

Once I have my overall laminate needs in hand, I then use a laminate planning sheet to plot the pieces. My planning sheet (see the drawing above) is nothing more than another ⅛-in. grid sheet (the same scale as the countertop diagram) on which I've scaled off a 5-ft. by 12-ft. sheet of laminate. Around the perimeter of the planning sheet I number the 1-ft. increments in feet and inches to simplify the figuring. If you adopt this system, save your original planning sheet and

photocopy worksheets as needed. When necessary, the 5x12 grid can be marked down to correspond to any other size sheet of laminate.

I use a pencil to draw in the pieces and erase frequently as I jockey lines around to get the most efficient layout. On this job my main objective was to make sure the one necessary seam ended up behind the stove. Once the pieces were arranged, I labeled each with a number and then labeled the countertop diagram accordingly (see the drawing on the facing page).

With a pencil and good eraser, plot the pieces on the worksheet. (In this example, all the pieces fit onto one 5x12 sheet of laminate. When the pieces are plotted, label each with the appropriate number to match the sheet layout diagram (see p. 31).

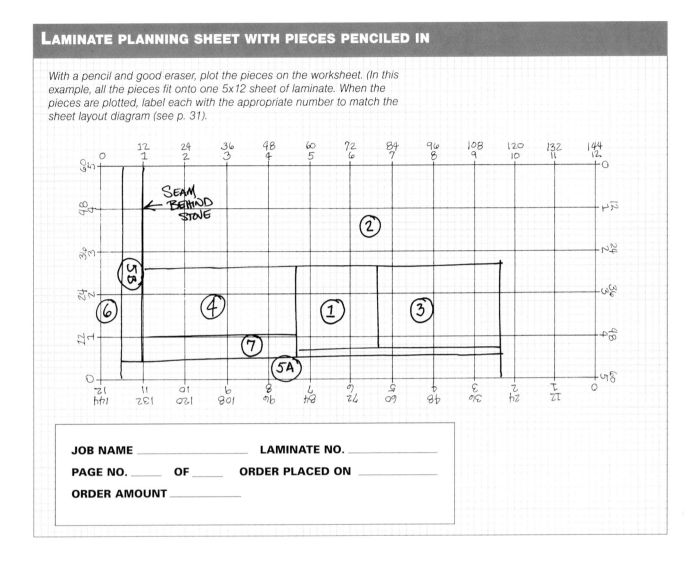

After the major components are plotted, I then utilize any substantial sections of extra laminate by figuring in the edge and backsplash top pieces. In this example, there was very little laminate left over, so I ordered enough rolls of self edge to do the job. It was much cheaper than buying another sheet of laminate.

Estimating the laminate for this countertop took me about an hour, and I was lucky—most tops don't fit so efficiently into one sheet of laminate. Whether you use this figuring system or another one of your own design, when you are done you will be able to order your laminate with confidence, knowing there will be enough and that the goods will be used efficiently. Then, when the material arrives, the drawings will guide you as you mark the sheets and cut out the pieces you need for the job.

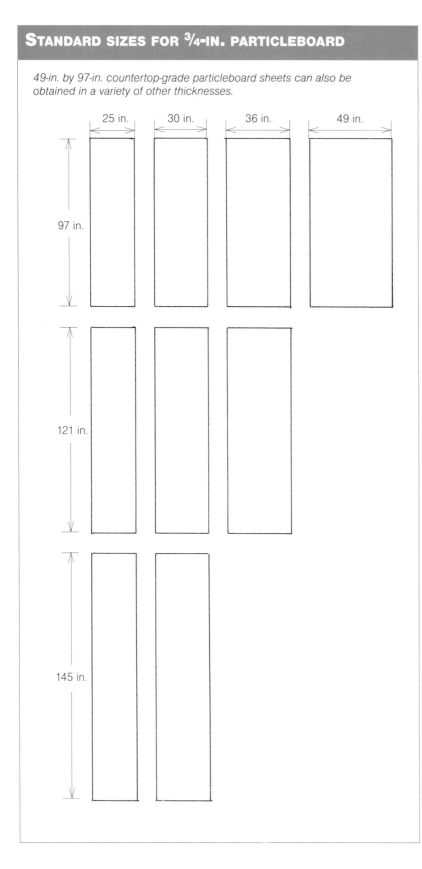

STANDARD SIZES FOR ¾-IN. PARTICLEBOARD

49-in. by 97-in. countertop-grade particleboard sheets can also be obtained in a variety of other thicknesses.

25 in. 30 in. 36 in. 49 in.

97 in.

121 in.

145 in.

Underlayment

Once you've calculated your laminate needs, figuring materials for the underlayment is a snap. Any lumber store or home center carries ¾-in. thick medium-density, countertop-grade particleboard in a variety of widths and lengths (see the drawing at left). It's likely that some, but not all, sizes will be in stock, so it's best to figure your underlayment needs when you do the laminate, and order the two together.

Sheets 25 in. wide are ideal for decking because in many instances, 25 in. will be an acceptable countertop depth. If that width doesn't quite give you enough front overhang, you can gain another ¾ in. by adding a facing strip (see the drawing on the facing page). If adding the strip makes the deck a skosh too deep, trim the back of the sheet to the desired finish size.

I frequently use 30-in. wide sheets of particleboard because I can rip a 4-in. wide length of backsplash off them and have 26-in. wide decking pieces left over, which is usually plenty for overhang in the front and scribing at the back wall. Most peninsulas and islands with extended overhangs can be taken care of with a single 4-ft. by 8-ft. sheet of deck material. (Like laminate sheets, particleboard sheets are also about an inch oversize in both dimensions, so a nominal 4x8 sheet is actually 49 in. by 97 in.)

Properly made underlayment seams can be placed almost anywhere, though it's best to avoid putting them in weak areas like sink and range-top cut-outs, and on extended overhangs. All seams in the decking should be solidly supported on the underside with a cleat at least 12 in. wide and centered on the seam.

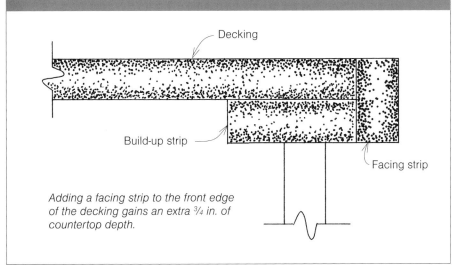

Decking

Build-up strip

Facing strip

Adding a facing strip to the front edge of the decking gains an extra ³/₄ in. of countertop depth.

Build-up strips should be at least 2 in. wide, but I prefer to make them about 3 in. wide in case I end up trimming some off the edge later. And the extra-wide build-up means there will be plenty of solid bearing on the top of the cabinets. When it comes time to attach the the countertop with screws driven up through the cabinet bracing into the underside of the counter, a wide build-up also provides ample material for the screws to find anchor.

If you end up with leftover particleboard at the end of a job, it can easily be used on the next countertop. And if by some chance you don't order quite enough, you can usually just zip down to the lumberyard for another sheet.

Adhesives

Contact adhesives are commonly available in quart-, gallon- and five-gallon containers. When calculating adhesive needs, first determine the total square footage of countertop, double the area (because you will be covering both the top of the underlayment and the under-side of the laminate sheet), and then compare that figure with the coverage guidelines on the can of adhesive you intend to use.

Water-based adhesives, which have a relatively thin consistency, will cover about three times as much area as solvent-based adhesives. Keep in mind also that coverage will vary somewhat depending on how heavily you apply the adhesive (see p. 74), and for brush-and-roller grades of glue, the manufacturer's estimated coverage will probably be a bit high. So it's usually wise to buy a bit more adhesive than the label says you'll need.

As a rule, a quart of solvent-based adhesive should be sufficient to glue up about 8 linear ft. of standard 2-ft. deep countertop (with backsplash), and a quart of water-based adhesive should do 27 linear ft.

3

ROUGH CONSTRUCTION

Now that you're familiar with the materials and components of a countertop and how they fit together into a coherent, functional design, you're almost ready to take out your tools and start building. In this chapter you'll learn about roughing out the countertop components: the deck, the build-up strips, the cleats and the laminate. But before you start, you must decide where you're going to perform the work, on site or in your shop.

SITE-BUILT VS. SHOP-BUILT COUNTERTOPS

There are definite advantages to building a countertop where it will be installed as opposed to somewhere else. The major advantage to site-building a countertop is that you can build the underlayment right on the cabinets and easily fit it precisely to any wall irregularities or a complicated cabinet layout. Building an involved layout apart from the cabinets requires careful measurements, and, when you finally fit it, adjustments (or compromises) usually have to be made.

Site-built countertops are also much less susceptible to dimensional movement problems than tops built in a shop, which may have radically different temperature and humidity from the kitchen where the top will be installed. For example, my own shop, which is heated in the winter only on days when I'm working there, does not provide a good environment for proper acclimation of materials before installation (see the discussion of temperature and humidity on pp. 9-12).

Some countertops are so large and complicated that site-building is the only practical option. If you build the top where it will be installed, you eliminate the hassle of transporting a bulky, unwieldy thing across town, up stairs, through doorways, and into position.

You've probably guessed by now that I am partial to site-built tops, and most of the countertops I fabricate are in fact built on site. But I would be remiss if I didn't point out that site-building has disadvantages and limitations too. Materials must be delivered to the site at least 48 hours before fabrication (or you can build the underlayment in place and then wait, to allow for proper acclimation, before gluing on the laminate), and the fabrication process creates a considerable amount of dust and mess in the customer's home or place of business. There are occasions—busy commercial settings, for example—when assembling a countertop on site is a logistical nightmare. But all in all, I think that the advantages of site-building outweigh the disadvantages.

In the shop-built construction method, build-up strips are arranged and fastened to the underside of the deck.

Whether you build in the shop or on site, there are various assembly methods and combinations of methods for arriving at the same finished product. For the sake of clarity, I divide these methods into two categories—shop built and site built. Please note that I said "methods" because the difference is not necessarily where the countertop is built, but how it is built.

The shop-built method entails turning the deck upside-down, then arranging and fastening the build-up pieces to the bottom of the sheet (see the photo above). Then the assembled underlayment is turned right side up, the laminate is applied, and the completed countertop is set into its final resting place. By this definition, it's not necessary to build a shop-built countertop in a shop; it could be built anywhere (even on site) apart from its final location. The site-built method entails building directly on top of the cabinets in a right-side-up position, with the build-up attached with screws driven down through the top of the deck.

FASTENERS AND FILLERS

Screws, staples and glue are used in various combinations to fasten decks, build-up strips and cleats to each other. On a site-built countertop, the holes left by the countersunk screws must be filled, and the surface smoothed, before the laminate sheet can be cemented down. On a shop-built countertop, the top remains free of screw-head imperfections, and only minimal surface preparation is needed before the laminate is applied.

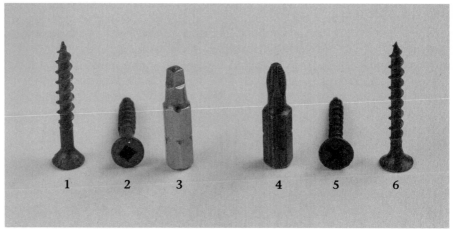

Particleboard screws (1 and 2), which are designed to take the stress of being driven without a pilot hole into particleboard, have thicker shanks and coarser threads than drywall screws (5 and 6). Square-drive screws (2) are easier to drive in than Phillips screws (5), and the bits used to drive them (3 and 4, respectively) will last at least twice as long.

Shop-built methods of construction

Using the shop-built method, I usually fasten the build-up to the deck with screws and glue, but some people glue and staple the two together. One fabricator I know dispenses with glue entirely and fastens his build-ups with plenty of 1¼-in. long narrow-crown (⁵⁄₁₆-in.) staples fired in with a pneumatic staple gun. He says he's never had a problem with anything coming loose. I've used a stapler too, and with good success, but I don't feel it's prudent to forgo the glue, particularly on the front-edge build-up and on cleats.

The thing I like about screws is that they pull everything together so tightly. Back in the old days (which, in my case, weren't really that long ago) I used to attach pine build-up strips with glue and straight-slot wood screws that I twisted in with a hand-operated screwdriver. To make the job easier and prevent the pine from splitting, I'd first drill a clearance hole and then a countersink for the head, but the process was still hard, time-consuming work.

When Phillips-tip drywall screws hit the market, I naturally adapted them to my purposes, and drove them home with a Phillips bit chucked in an electric drill. With all that new driving power, I discovered the pilot and countersink weren't always necessary, but the pine would still sometimes split.

When I switched to a particleboard build-up, there was never a problem with splitting, but the drywall screws often couldn't take the torque needed to twist into two layers of the dense material and the heads would twist off. Sometimes the threads wouldn't grab sufficiently, and the screws would spin uselessly in place. So it was one step forward and one step back until I discovered particleboard screws. These little gems look like drywall screws but have a high thread profile and are stout enough to handle the stress.

Particleboard screws (see the photo above) are widely available from several suppliers, but the ones I use exclusively for attaching my build-ups are the McFeely's #8 x 1¼-in. flat-head steel

screws with nibs. These are slight cutting protrusions on the underside of the head that help it countersink itself in the high-density surface of a piece of particleboard. The screws I use are so aggressive and tough that they'll fasten a particleboard build-up on without a pilot hole, and countersink themselves as much as ¼ in. without the slightest hesitation. I've never had one snap.

While most particleboard screws have the usual #2 Phillips drive head, the McFeely's screws have a #2 square drive. Square drives have been around since 1908, but they're just starting to

become popular with woodworkers. If you've never used a square drive, you should try it, especially for this application. A square-drive bit in an electric drill will muscle in screws with considerably less effort on your part, and the bits last much longer than Phillips tips.

In conjunction with the screws, I use yellow wood glue and apply it by holding the bottle (the kind with a dispensing spout) above the build-up piece and drizzling on an ample amount (see the photo above). Sometimes I'll spread it out, but usually I don't. When the assembly is screwed down, a neat little

When assembling a shop-built underlayment, space screws down the center of the build-up.

line of glue oozing out between the two pieces indicates a good job of gluing, but copious gobs of excess adhesive mean you've been too liberal with the glue bottle. Space the screws down the center of the build-up strip at 6 in. to 8 in. apart (see the photo above), or as close as you feel is necessary to pull the pieces together and seat the glue. (It is the glue more than the screws that will hold an underlayment assembly together. In fact, on a couple of jobs I've removed all the screws after the glue dried and reused them.)

Site-built methods of construction

With a site-built countertop, the build-up strips are first positioned on top of the cabinets (see the photo at left), then the deck sheets are placed on top of the strips. If need be, the strips can be held in place with a few spots of hot-melt

In the site-built construction method, the build-up is arranged on top of the cabinets before the deck is glued on.

glue or tape. Easy-to-reach build-up pieces (like the ones on the front edges) may be slid in place after the deck is down. The parts are fastened together with McFeely's screws driven down through the top of the deck. The countersunk screw heads will create a pockmarked top surface, so the imperfections must be filled before the laminate is applied. Because of this need to fill the holes and position the strips from underneath, site-built countertops require a little more fabrication work than shop-built tops, but the work isn't difficult and needn't take much more time, especially when you look at the total picture and consider that site-built tops are form fit in place, whereas shop-built tops need to be installed.

I recommend two-part auto-body filler for filling the screw heads, as well as any gaps or other imperfections in the underlayment; it mixes easily, dries fast and hard, and sands relatively well. Auto-body filler is available in any auto-supply store and in most department-store auto departments. It comes as a can of light-colored paste-consistency resin and includes a small tube of dark red hardener.

It doesn't take much hardener to set the resin. The exact proportion of hardener to resin is not critical, but the more hardener used, the faster the filler will set up. I mix the two instinctively and somewhat arbitrarily. If you've never done it before, follow the mixing instructions on the can. If there are no mixing instructions, start by squeezing a ½-in. to ¾-in. length of hardener onto a glob of resin approximately 2 in. in diameter. Mix the two ingredients using a 2-in. to 3-in. wide putty knife on a scrap of laminate, a piece of wood, or anything clean, smooth and disposable (see the photo above right). Mix until the resin is uniformly colored by the hardener. Then, without further ado, fill the imperfections (see the photo at right), and when this is done, it's best to overfill a bit.

Two-part auto-body filler sets up quick and hard, and is ideal for filling and repairing imperfections in the underlayment. Only a small amount of hardener is needed to set the resin.

Work the filler into the holes with a putty knife.

Use a sanding block with 36-grit sandpaper to remove excess body filler.

A sharp block plane does a good job of smoothing filler and leveling sections of decking that don't meet evenly.

In no time the filler will start to get rubbery, then harden up rock solid. When it has become sufficiently firm, and, preferably, before it gets rock hard, remove the excess filler. For this you can use a sanding block outfitted with coarse sandpaper (I like 36 grit) or a sharp block plane followed by the sandpaper to smooth down any ridges, as shown in the photos above. A belt sander will also do a good job at smoothing a filled surface. However, be aware that belt sanders can be overly aggressive; make sure to hold the

sander's base flat to the surface, and keep the tool in constant motion when operating. Be especially careful when belt sanding close to the edges—gouges there are really bad news. If one coat of filler doesn't adequately fill the depressions, go through the process again where needed.

CUTTING THE DECK TO SIZE

At some point in the underlayment construction process, you'll have to cut the underlayment to exact finished size. This

If the deck is cut to finished size before the build-up is attached, use a square to flush up the edge of the build-up with the deck as you fasten it down.

can be done either before or after the build-up is attached. If the deck is precut to finished size before the strips are attached, they must be flush fit. That is, the build-up must be aligned flush with the final cut edge of the deck (preferably using a square to ensure accuracy) before it is fastened on (see the photo above).

The alternative cutting approach, and the one I prefer to use if possible, is to rough-cut the deck about 1 in. oversize in each direction, fasten the build-up on without regard to exact flush alignment, (see the photo at right) and then cut the top to precise finished size after assembly (a build-up strip as narrow as 2 in. will still provide adequate support). I have found this assemble-then-cut method is easier and faster to do, and it renders truer edges for gluing to. If you use this method and have attached your

If the deck is cut approximately 1 in. oversize, you can attach the build-up strips without regard to precise edge alignment, and then use a circular saw to cut the assembly to finished size. Here, a length of MDF acts as a straightedge to guide the saw.

build-up with a row of screws down the center, they should be well out of the way of the sawblade. But it's still a good idea to double-check and remove fasteners if they are in the line of cut, since they are hard enough to ruin any blade.

Assemble-then-cut underlayment production is suitable for any countertop, even site-built, provided the top can be pulled away from the wall enough to allow the saw room to do its work. For site-built configurations that will not allow movement, you will have to use the precut/flush-fit method.

WHY SQUARE-CUT EDGES ARE IMPORTANT

Later in the fabrication process, after the countertop's side and top pieces of laminate have been installed, the overhanging top sheet will need to be trimmed flush with the edges. This task is typically done with a flush-cutting router bit guided by a bottom-mounted pilot

bearing. Therefore the underlayment must have a perfectly square edge cut, because even a slight angle would cause the bearing to mis-guide the bit to take off either too much or not enough (see the drawing below).

Square edge guide, accurate flush cut

The dotted lines indicate the cut of a flush-trimming router bit as it contacts a square edge (top), as opposed to out-of-square edges (center and bottom).

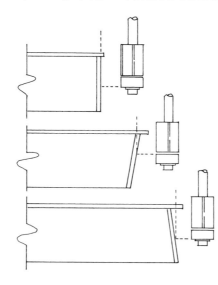

Bearing rides against square edge; bit trims laminate sheet flush with edge strip.

Bearing rides against outward-sloping edge; bit snipes off upper part of edge strip.

Bearing rides against inward-sloping edge; bit doesn't trim laminate sheet enough.

Cutting straight underlayment edges

All well-made underlayments must have straight and square finished edges (see the sidebar on the facing page), and the key to making straight cuts is to use a circular saw guided along a straight-edge. Freehand cuts do not a straight line make, and a portable circular saw is the only portable power tool that can make a square edge cut. (Save your jigsaws and reciprocating saws for sink and range-top cutouts and for coping the underlayment to the back wall. They are not well suited for making square edges.)

For finished edge cuts I use my trusty Porter-Cable 7¼-in. circular saw. My setup for cutting any kind of sheet goods is always the same: two sawhorses set 4 ft. to 5 ft. apart, with three spaced-apart 2x4s laid across the tops. This arrangement will support just about any size sheet on both sides of the cut line. With 2x4 support pieces I can rip or crosscut a sheet without fighting to keep it from flexing and the saw from binding in the kerf. I set my blade just deep enough to slice through the sheet material; little kerf tracks on the supports don't bother me, but through cuts are obviously not desirable.

A simple width gauge lets me position the straightedge to guide my saw. The gauge can be made out of scrap material so it doesn't cost anything, it saves time, and it substantially reduces the chance of making a measuring error. My width gauge (see the drawing above right) is made from a piece of plastic laminate. It is 8 in. long and as wide as the distance between my saw's base-plate edge and the blade's cut line. This distance varies from saw to saw; the baseplate of my Porter-Cable is 8½ in. wide but as is usually the case, my blade is positioned well off to one side. For accurate cuts I run the wider part of the base with full bearing on the sheet, and along the straightedge. On my saw, the

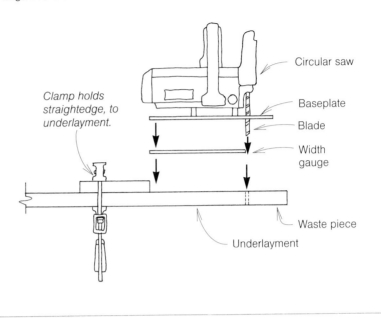

SIZING A WIDTH GAUGE

Make your gauge as wide as the distance between the baseplate of your saw and the edge of the blade.

Circular saw

Clamp holds straightedge, to underlayment.

Baseplate

Blade

Width gauge

Waste piece

Underlayment

distance from the wide side of the base to the cut line is 6¹¹⁄₁₆ in., and that's the critical measurement of my gauge.

To use the gauge, I hold it to the marks on my underlayment sheet where I've determined that I want the cut, and mark the opposite side of the gauge, which is where the straightedge will be clamped down (see the drawing on p. 46). A small mark on each end of the sheet is all that's needed.

To avoid confusion, I always clamp my straightedge to the piece of material I intend to keep. If the gauge and guide were positioned on the waste side of the line, the "keeper" would be a blade kerf too small (try it and you'll see what I mean).

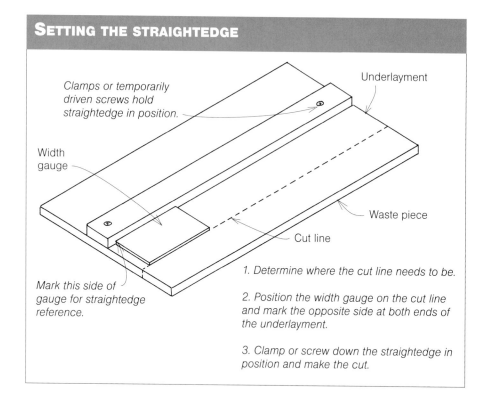

SETTING THE STRAIGHTEDGE

Clamps or temporarily driven screws hold straightedge in position.

Underlayment

Width gauge

Waste piece

Cut line

Mark this side of gauge for straightedge reference.

1. Determine where the cut line needs to be.

2. Position the width gauge on the cut line and mark the opposite side at both ends of the underlayment.

3. Clamp or screw down the straightedge in position and make the cut.

For my straightedge guides, depending on the distance I'll be cutting, I use either my 4-ft. clamp-and-guide straightedge or the manufactured edge of an 8-in. wide 97-in. strip of MDF that I keep around specifically for a straightedge. I also have a similar straight length of particleboard that's 10 ft. long but I seldom use it.

The nice thing about a wide strip of MDF or particleboard straightedge is that you can drive two or three temporary McFeely screws through to hold it in place, instead of using clamps on the ends. And the wood edges are surprisingly durable; I used one 97-in. edge for three years before cutting it into build-up strips.

Before directing the blade into good material, it's wise to make a test cut to check for squareness. I once cut a large section of underlayment to size and discovered when I was done that I hadn't reset my saw base after making a slight bevel cut on a door edge the day before.

As for sawblades, use ones with carbide teeth. Particleboard is tough stuff, and ordinary steel blades won't cut the proverbial mustard. The more teeth on the blade, the smoother the cut. I use a 24-tooth, thin-kerf Irwin Marathon blade on my saw. It's relatively inexpensive, and I think it holds a good edge remarkably longer than other carbide blades I've used.

ATTACHING CLEATS AND LEDGERS

When two pieces of underlayment are butted together, the underside of the seam should be supported along its full length by a cleat, or a combination of cleat in the center and build-up on the edges (see the top photo on the facing page). Always use glue and lots of fasteners when attaching cleats. The two abutting deck pieces may be joined with the aid of biscuits (see the bottom photo at right), but they don't have to be; glue and screws with a sufficiently sized cleat (minimum 12 in. wide, and centered under the seam) will do just fine. Occasionally, the joint may not end up perfectly tight, or the surface level between the two sheets may be slightly uneven. You can fix things by puttying the crack and by belt-sanding or block-planing down any unevenness.

It may be necessary to install wood ledger strips along the wall to support the finished countertop in areas where there is no cabinet to do the job. For example, lazy Susan corner base cabinets are often smaller than the space allowed for their installation. As a result, the underlayment is unsupported along the back wall, so you'll need to nail or screw a 2x4 ledger there. Some fabricators add a ledger strip to the back wall where a countertop spans the dishwasher opening. Another area that can benefit from a small support strip is under the end of a deep countertop overhang that abuts a wall.

FORMING RADIUSED CORNERS

Countertops with rounded corners are a nice design touch (see pp. 114-116), and they have practical advantages too. Pointy, squared-off outside corners are a painful head bump waiting to happen for the heedless toddler; rounded corners are less dangerous. Also, squared outside corners have a seam where the two edge pieces of laminate meet; they can be damaged and come loose if banged into with a hard object. Rounded corners have no corner seam.

The trick to properly cutting a curved corner in the underlayment is to make the radius symmetrical and the cut edge square with the top surface. (Such cuts are usually done after the build-up is on, and after the underlayment is cut to finished size. A corner with a large radius may require a larger build-up in that

Here a seam in the deck will be supported by build-up strips on the front and back and a wide cleat in between. Lots of glue and screws make a solid connection.

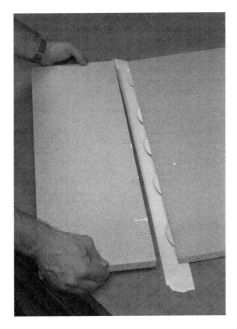

Biscuits may be used to help align adjoining sections of decking. If you use them, glue the particleboard edges; gluing the biscuits themselves is optional because the cleat will make the connection strong enough. Masking tape underneath will keep the glue ooze from getting on the work table.

area.) One method for making the cut is to mark the desired radius on the underlayment, then cut the curve with a jigsaw. The problem with this approach, though, is that I've never seen a jigsaw blade follow a curve in 1½-in. thick material and not wander off course at the bottom, resulting in an out-of-square cut. It's possible to make the cut oversize, and then rasp, sand or plane up to the line using a square over the edge as a guide for accuracy, but this is tedious work. Nevertheless, hand-shaped curves are possible, and I've made many of them over the years.

However, hand-cut curves will almost always fall short of perfection; the laminate edge will probably not fit tight to the underlayment in all places as it makes the bend. There will usually be a gap at the top or bottom, and the laminate will be "spongy" in that spot. If this flexible area is on the top edge where a trimmer-bit guide bearing will ride when

trimming the top sheet of laminate, the bearing will roll into the soft spot and end up cutting off too much laminate. The problem can be avoided by filling the gaps with auto-body filler (see p. 41).

It's possible to buy adjustable corner-rounding attachments for the router that will quickly trim off a precise radius (see Resources on pp. 127-129), but they are an expensive investment for such a dedicated purpose, and the range of sizes they can make is limited. Fortunately, there is a way to machine crisp curves of any shape or size, with minimal investment: a premade template and a router equipped with a bearing-guided straight bit.

Template routing a curved corner

Begin by making a template of the corner radius you want. I make mine out of ¼-in. Masonite and keep several on hand (see the photo at left). If the need arises, I can quickly make any other profile I need by drawing the outline on the Masonite, jigsawing slightly oversize, and carefully sanding up to the line. The ¼-in. Masonite is easily shaped and, once formed, quite durable.

To use, position the template in place where you want the cut, then jigsaw the bulk of the waste away, leaving approximately ½ in. extra. Now set the router base on the template, with the shank-mounted bearing positioned to ride against the template edge, and make the final cut. It will correspond precisely to the curve of the template, and the edge will be perfectly square all the way around (see the drawing on the facing page).

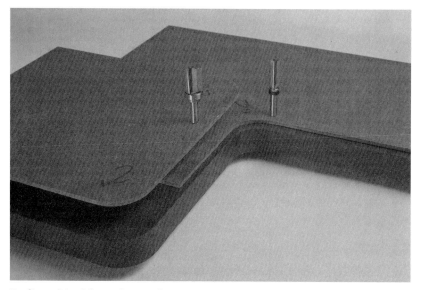

Radiused inside and outside corners in a countertop can be made with Masonite templates and bearing-guided straight bits.

MAKING A CURVED CORNER

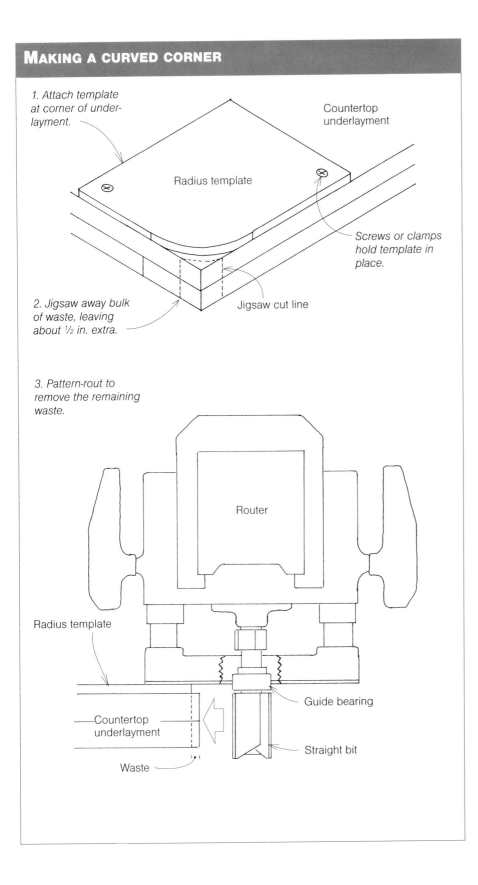

1. Attach template at corner of under-layment.

Countertop underlayment

Radius template

Screws or clamps hold template in place.

2. Jigsaw away bulk of waste, leaving about ½ in. extra.

Jigsaw cut line

3. Pattern-rout to remove the remaining waste.

Router

Radius template

Guide bearing

Countertop underlayment

Straight bit

Waste

To make a sink cutout in the underlayment before the laminate is applied, drill a hole at each corner of the layout, then jigsaw the front, back and side cuts. A scrapwood cleat screwed to the cutout will keep it from dropping down when the side cuts are complete.

A flush-cutting straight bit with a bottom-mounted bearing can also be used if the template is fastened to the bottom of the underlayment. In either case, a full-sized router (preferably a plunge router) and a ½-in. dia. router bit with a cutting edge at least 1½ in. long make the best template-cutting combination, though other routers and bit sizes can be used as well.

MAKING SINK AND STOVETOP CUTOUTS

Cutouts for a sink, stovetop or other appliance can be made before or after the laminate is applied. I prefer to make the cut in the underlayment before gluing down the laminate (see the photo above). I came to this conclusion after once cutting the opening for a stovetop unit in a finished counter and finding I had made it a tad too big. The top was a large U-shaped one that my helper and I had just spent two days building, and the incident stands as the most heinous mistake of my laminate career. I vowed it would never happen again.

Making the cutout first allows you to check the fit of the appliance. If, at that point, you find you've goofed, it's no problem to modify the opening and correct the mistake—cut the opening wider if it's too small, or glue on extra wood if it's too large. After the laminate is glued down over the opening, you simply rout out the superfluous material (see the photo on the facing page). Otherwise, there is little that can be done to rectify a botched cut in a finished counter (although, in the case of my stovetop, we had a thin chrome sub-rim custom made to cover the oversize cut, and fit the stove into it).

With a premade sink cutout in the underlayment, you simply rout out the excess laminate after the sheet is glued down. *Photo by Kevin Ireton.*

If you have to make a cutout in a finished top, remember the adage, measure twice (maybe even thrice) and cut once. When you make the cut, I recommend first drilling holes in each corner, then using a jigsaw with a reverse-tooth blade to make the long straight cuts. The average jigsaw blade has upward-angled teeth that cut on the upward stroke, and this arrangement could pull a small chip out of the good edge of the laminate. A reverse-tooth blade cuts on the downward stroke, and there is virtually no risk of chipping. (However, reverse-tooth blades do not work well with jigsaws in orbital action.) If a reverse-tooth blade is nowhere to be had when you need to make a cut, there shouldn't be a problem if you use a regular 12-tooth blade, keep the motor's rpm high, and cut slowly.

Another way to make cutouts, in either finished or unfinished tops, is to make a template of the sink and use a template-cutting bit in your router. Provided the template is sized right, this method ensures easily made, accurate and smooth-cut openings. Once made, a template can be used over and over (for sinks of the same size, of course). Proceed as described on pp. 49-50 for the template cutting of curves. Position the pattern and roughly jigsaw the opening out, leaving about ½ in. of extra material. Then rout out the rest of the waste.

Most new sinks and countertop inserts will come with either a sized diagram of the appropriate cutout or a full-size paper pattern. Patterns are particularly useful for round or oval sinks, but in lieu of a supplied pattern, you can make

A standard kitchen sink measures 22 in. front to back. For the sink to fit properly in a counter over 24-in. deep cabinets, it must be positioned as far forward as possible. Making the cutout ⅜ in. inside the cabinet face frame will typically leave about ½ in. between the backsplash and the back rim of the sink and still leave enough room for installing sink clips in the front.

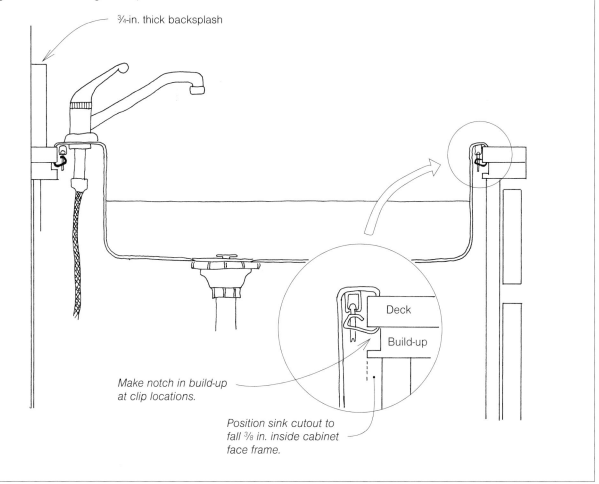

¾-in. thick backsplash

Deck

Build-up

Make notch in build-up
at clip locations.

Position sink cutout to
fall ⅜ in. inside cabinet
face frame.

your own by placing the sink, rim down, on a piece of cardboard, tracing the outline, and making the pattern approximately ¾ in. smaller (for porcelain and cast-iron self-rimming sinks) than the outside rim. Patterns aren't necessary for squared-off fixtures because you can easily make the measurements using the supplied diagram and a framing square. If no diagram is included, just take measurements right off the unit.

Never size cutouts for a tight fit because the countertop needs room for seasonal dimensional movement. Rims on items designed to be inset into a countertop will be sized to accommodate slightly oversize cuts. With most metal rims, ¼ in. of supporting material all around the rim provides adequate bearing and allows enough room for movement. However, as with any of my recommendations, always defer to a manufacturer's dimensions and suggestions, if you have them.

Where you will position kitchen sinks in relation to the depth of the counter is an important consideration. I always make sure to position the units as far toward the front edge of the counter as is possible, because most sinks are 22 in. deep and I want to have space at the back of the sink, especially if there will be an attached ¾-in. thick backsplash. The exact placement is determined by where the front frame of the cabinet falls. I place the front of the cutout ⅜ in. inside the cabinet face frame (see the drawing on the facing page), and this leaves just enough room for sink clips to grip the underlayment. Even if the sink doesn't require clips, this measurement is still valid.

The clips used to hold sinks tight to a counter (if the sink is a clip-attached type) are fastened on and tightened down after the sink is set in place, and the typical sink clip is designed to grab onto approximately ¾ in.—not coincidentally, also the thickness of the deck material. But in clip areas where the counter is 1½ in. thick (because of the build-up), the extra thickness has to be removed. I do this by drilling a short way into the center edge of the bottom build-up with a ⅞-in. spade bit, and I like to do this task while I'm building the underlayment—note the drilled clip pockets in the photo on p. 51. However, sink clips come in a fascinating range of styles (I even have a collection), and it seems that more sinks are using clips that have the capacity for a thicker (1½-in.) grip, so check out your clip situation before going to the trouble of making notches.

One very important thing when making any kind of cutout is to radius the corners. Sharp, square-cut inside corners are an invitation to stress cracking. Most fixtures will have a corner radius that

you should conform to anyway, but even a square-cornered rim should have at least a ⅛-in. radius, and ¼ in. is even better if possible. Such tight curves are best made with a drill bit.

One final note on cutouts: years ago I had a customer who let a pot boil over on her new counter's range top. The hot water seeped under the rim (most range tops do not clip tight to the counter) and into the underlayment, causing it to swell. The tremendous pressure from expanding wood cells caused a split several inches long in the laminate surface. It was a freak accident, but since then my standard operating procedure has been to seal the edges of range-top cutouts with a spread-on layer of caulk. It's also a great way to use up those old, partially full tubes.

TECHNIQUES FOR ROUGH-CUTTING LAMINATE

Once the underlayment is made, it's time to turn your attention to plastic laminate. The first task is rough-cutting the laminate. Rough cutting means cutting the laminate pieces bigger than their finish dimension; this is done to provide extra material for scribing, as well as leeway for positioning the sheet during glue-down. Later, the pieces of laminate are trimmed flush.

How much oversize you cut your laminate pieces is up to you. On large pieces, I aim for 1 in. extra all around. Edge strips are more manageable, and ¼ in. to ½ in. oversize all around is usually adequate. In practice, though, I usually end up with more overhang, and occasionally less. The point is that there must be at least some overhang.

Laminates can be rough-cut using general-purpose or specialized tools and machines. The most common methods are describedon pp. 54-59. All you really need for this work is an inexpensive carbide scoring tool. If you own a table

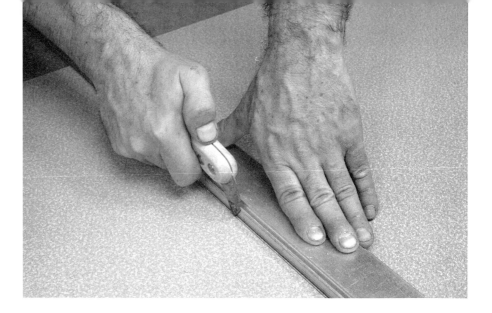

A carbide scoring tool is an inexpensive but effective implement for cutting laminate. The carbide tooth is guided along a straightedge and scores a groove through the decorative top layer of the laminate.

With the scored laminate face up, snap it by holding it down on one side of the line while lifting on the other side. It will break evenly along the scored mark.

saw, I recommend you make use of a Laminatrol guide. Hand shears are the next logical tool upgrade. If you really want to be well equipped, get a hand slitter and some electric shears.

Carbide scoring tool

A carbide scoring tool is nothing more that one carbide tooth mounted on a handle. It is an inexpensive (about $10) and sufficient tool for making any kind of a laminate rough cut, short of curves. You use it to score a line in the laminate along a straightedge (see the photo above), and then snap the laminate sheet along the line (see the photo at left).

One or two passes of the tool will score a groove through the decorative layer and far enough into the phenolic backer to allow the piece to be snapped apart. Snapping is done by holding down firmly on one side of the line and lifting the laminate up on the other. There will be some resistance, but the piece will suddenly snap cleanly along the line. Don't bend the scored sheet down to make the snap because it will not break cleanly. People who have never used a scoring tool before will typically score

When glued to a countertop underlayment, plastic laminate makes a durable finished surface, but in sheet form it can be brittle. Nowhere is this more of a problem than when you are making inside-corner cutouts.

If you rough-cut a sheet with an inside corner, it is best to stop your cuts short of the inside corner point and connect the two cut lines at an approximate 45° angle (or make a wide rounded cut if you have the tool to do it). Making this kind of cut is more difficult but it provides more strength in the corner. Nevertheless, the corner remains fragile, and anyone who lifts or otherwise moves such a sheet should focus on that inside corner to make sure that it doesn't undergo any twisting stress.

Also, if you're leaning over a cutout corner to scribe the back to a wall, don't lean too close. I've ruined more than one sheet (not to mention my day), by not being mindful of the weak inside corner.

many times and much deeper than they need to, but, aside from being more work, that's not a problem because the material will just snap more easily. Sometimes it is necessary to score all the way through, as when cutting a square notch out of a sheet; one length of the notch would need to be sliced through, and the other could then be scored and snapped. For another way to handle inside-corner cuts, see the sidebar above. If you've never used a carbide scoring tool, it's a good idea to practice making cuts on scrap laminate before tackling the big project.

I used a scoring tool exclusively for several years of laminate cutting, and still keep it close by, but almost never reach for it any more. The problem is that to cut big sheets, you need a large, clean flat surface to lay the sheet on for scoring, and that isn't always convenient (see the sidebar at right). Besides, scoring is labor intensive, especially when there are lots of long strips to cut. And if you aren't careful, the carbide tooth will wander away from the straightedge before you realize it. For anyone who is going to be cutting a lot of laminate, there are better tools.

Sometimes it's necessary (or it just makes work easier) to lay a large sheet of laminate down to mark out and cut pieces. But finding a spot that's flat, hard, clean, and big enough to lay out and work on a 5x12 sheet of laminate is often difficult, especially if you're site building the top. A garage or patio might be big enough, but garage floors typically have oil spots that could transfer to the back of the sheet and inhibit the glue bond. And most garage and patio floors are either not smooth or difficult to sweep clean. Once I laid a sheet of laminate on a garage floor to make a score and snap cut. I had swept the spot meticulously, I thought, but the downward pressure of the cut caused a tiny stone I had missed to crack the laminate surface. The stone was minuscule, almost dust, but it was hard and unyielding, and the laminate was no match for it. I should have known better, but I was desperate for space.

Now, when necessary, I make my cuts on clean sheets of ¼-in. lauan plywood or Masonite. I don't consider these an extra expense because I regularly use them layered over dropcloths to protect delicate floors when I'm working over them. For a laminate-cutting work surface, I lay out three sheets, making an 8x12 working surface, which is ideal. You may not want to go to that expense, but even if you get just one sheet of plywood and keep it under where you are cutting, it will make the job easier. You'll still want to give the sheets a good sweeping, but they'll come clean.

Keep in mind, while you're working with a sheet on the floor, that any hard objects that fall on the sheet (like that tape measure you thought was clipped securely to your belt) may damage the laminate too.

A Laminatrol guide is the perfect table-saw accessory for keeping laminate from sliding under the fence or lifting up when ripping strips.

Table saw

A table saw makes quick work of rough-cutting strips of laminate. For chip-free cutting, laminate manufacturers recommend a 60-tooth, 10-in. carbide blade with a tooth pitch of .417 in. and a blade rake angle of 10° to 15°. I'm not that particular; minor chipping in rough-cut pieces of laminate doesn't bother me, and I've used various carbide-tooth blades with good results. A lot of teeth and a slow feed rate will yield the smoothest cut. Always run the laminate through the saw good side up.

One problem with cutting laminates on a table saw is that the thin material tends either to lift as it's being cut or to slide under the fence. The simplest way to circumvent both of these problems is to use a Laminatrol guide (see Resources on pp. 127-129). This inexpensive aluminum extrusion (see the photo above) fits under the saw fence and has a groove along its length that guides and holds down the fence side of the laminate.

There are other disadvantages to table-saw cutting. First, of course, you need a table saw (something not always available or convenient to take to a job site). You also usually need a helper to help guide unwieldy sheets through the saw. Finally, a table saw is not appropriate for cuts on large pieces or for making odd shapes.

Laminate hand shears

Laminate hand shears, or snips, are one of my favorite tools for cutting laminate. Laminate snips have a center cutting blade that pivots between two other blades and removes a ⅛-in. strip of laminate from the cut line (see the top photo on the facing page). It's possible to cut laminate with regular tin snips, but it's a risky endeavor and not recommended, because the cut can suddenly crack off course. With laminate snips there is little chance of uncontrolled cracking.

Hand shears are available in two straight-cutting models; one is held above the sheet when cutting, and the other below. A third style (called universal) does not work efficiently for straight cutting, but does let you cut curves (see the bottom photo on the facing page). The tool doesn't actually cut a curve, but each squeeze of the handles punches out ⁵⁄₁₆-in. by ½-in. pieces of laminate; by overlapping the punch-outs, the tool can be used to cut a decent rough curve.

Laminate can be shear-cut while resting in a vertical position (leaning against a wall) or supported horizontally across sawhorses and 2x4s; there is no need for a large flat cutting surface. And the shears, though more expensive than a scoring tool, are still relatively inexpensive—about $35. The biggest drawback to laminate snips is that they can be tiring to use; the squeezing muscles of your hand and lower arm will undoubtedly let you know this as you make your way through a long cut. But once I got

Laminate snips allow for controlled snip-cutting by shearing out a ⅛-in. strip of laminate.

used to using the snips, I didn't find them all that difficult to operate. Eventually the spring in the handle will weaken, making the cutting even more challenging, but replacement springs are available.

Electric hand shears

If you start cutting laminate with a carbide scoring tool, then move up to hand shears and like the cutting versatility but not the hard work, you'll love motorized hand shears. They work with the same center-pivot, shear-cutting action, but the model I use (see the photo on p. 58) cuts a ¼-in. kerf.

Squeezing an on/off trigger switch is a lot easier and faster than squeezing the handles of hand shears repeatedly to make a long laminate cut. Electric shears will also cut gradual curves that would be otherwise difficult with manual

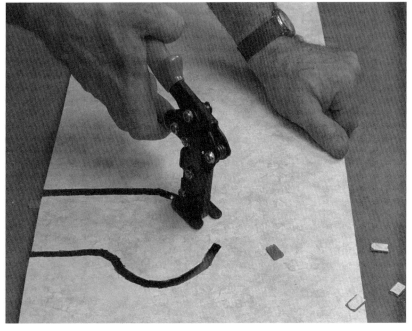

Universal-style snips punch out little pieces of laminate. This tool is not efficient for straight cutting, but will do what the other styles won't—cut curves.

Electric shears work quickly and are easier to use than manual shears. They will also cut gradual curves.

Hand slitter

A hand slitter is a useful tool for cutting laminate strips out of a sheet. Unlike the table saw, the slitter is small and portable, and it doesn't make noise or dust when it cuts. The slitter (see the photo at left on the facing page) is operated by pushing or pulling it along the edge of a sheet of laminate. Two small cutting wheels pinch and lightly score the top and bottom of the laminate as it passes through. This action weakens the laminate sufficiently for it to be easily snapped apart. The cutting wheels are adjustable and can be positioned to shear completely through the laminate, but the increased resistance makes it harder to operate and control the tool.

Width of cut is determined by an integral side guide that can be adjusted for cuts from $\frac{1}{4}$ in. to $3\frac{1}{4}$ in. Since $3\frac{1}{4}$ in. is not enough for rough-cutting back-splash pieces, many fabricators outfit their slitters with optional extension cutters that increase the potential cut up to $4\frac{1}{16}$ in. But with the extensions, the narrowest cut possible then becomes $\frac{9}{16}$ in. Another option for the slitter is a table clamp that holds the tool stationary while sheets are fed through the cutter.

Unlike any of the other cutting methods, hand-slitting wastes no laminate in the process of making a cut. The cut edges are relatively clean, though sometimes there is some minor splitting. Once you get the hang of using it, a hand slitter is a handy tool to have. But it isn't necessarily trouble free: if you're not careful, the tool will wander off course, and it does require a fair amount of effort to push through the laminate. Also, a slitter, like a table saw, requires that the laminate have one straight edge to index the cuts.

Motorized slitter

Porter-Cable is the only company that makes a motorized slitter. It can be purchased as an individual tool or as a sub-base attachment for the Porter-

shears. Although the cutting action is aggressive, the speed and direction of cut are not too hard to control. However, the shearing action of these shears creates a coarser cut that is much more susceptible to minor chipping on the edge than a cut made with manual shears.

Some electric shears have adjustable edge guides, which I imagine would be a nice feature to have. And it is possible to buy a laminate shear attachment for your $\frac{3}{8}$-in. electric drill. These are available through laminate tool supply catalogs (see Resources on pp. 127-129).

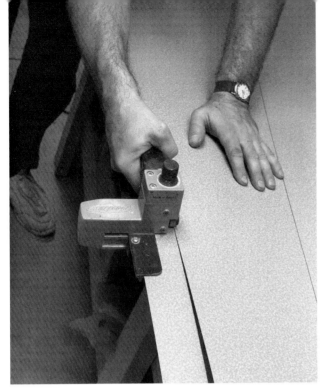

A hand slitter is a portable, quiet and dust-free way to cut strips of laminate. It has an integral side guide and can be pushed or pulled through the cut.

This motorized slitter comes as an attachment for Porter-Cable's #7310 laminate trimmer and will cut smooth, accurately machined strips of laminate.

Cable #7310 laminate trimmer (see the photo above right). Rather than score or shear off strips of plastic, the Porter-Cable slitter has a $\frac{3}{16}$-in. carbide-tipped straight router bit that cuts through the laminate. The sharp, small-diameter cutter does its job with ease and produces smooth and accurately machined strips. (Such precision is not necessary when rough-cutting laminate, but would prove very useful for making inlay strips in a countertop edge.)

The slitter's cutting bit is self-contained, which means you could lay the sheet of laminate being cut on any surface (like carpeting), run the machine down an edge, and not worry about the bit damaging the underneath material. Cuts to 4¼ in. wide are possible.

I once owned a Porter-Cable slitter and almost never used it (I ended up selling it because I found it awkward to operate), but I had occasion to try another one a few years later and it worked beautifully.

I think the difference was that on mine I had used a larger-diameter bit that wasn't very sharp. As far as drawbacks to the tool, it emits a high-pitched scream, like any router, and makes a lot of dust. Also, since it's a motorized tool, it's more expensive than a hand tool— about $200.

Router and bearing-guided straight bit

Another way to make rough sizing cuts in plastic laminate is with a router and a bearing-guided straight bit. If you position the laminate over a straightedge (such as the 8-ft. length of MDF I use) and clamp it in position, the router bit can be directed through the laminate, and the bit's bearing will ride along the straightedge (see the photo at right). The process is similar to template cutting, with the straightedge acting as the template guide. This method of cutting strips is simple, inexpensive and entirely adequate for most cuts.

A straightedge and bearing-guided straight bit in the router can be a useful combination for rough-cutting jobs.

4

GETTING A GOOD FIT

The process of fabricating a plastic-laminate countertop can be divided into two phases, the rough phase and the delicate phase. The rough phase, discussed in Chapter 3, encompasses the underlayment construction and oversize cutting of the laminate pieces. If you make a mistake when assembling the underlayment, there is no great problem because you can mend it with some body filler or by adding on a piece of wood. Rough cutting the laminate pieces may be a bit tricky, but it is still only rough cutting and there's some room for error. The delicate phase, which is the subject of this chapter, involves fitting the laminated top to the wall and seaming adjoining pieces of laminate. In these operations there is almost no margin for error; even tiny mistakes can become visible (and maybe even glaring) flaws.

FITTING THE COUNTERTOP TO THE WALL

One of the great differences between a poorly built or average countertop and an exceptionally well-crafted one is how well it fits against the wall and anything else it meets. If the counter is a straight section against a straight wall, one would be hard pressed not to make a good fit, but what if the countertop fits against two out-of-square walls or three walls? What if it jogs into a bay window or butts against an out-of-square wood

post, or ends abruptly at a finished wood panel and there will be no molding to cover it? To handle these tricky situations you'll need to master four basic laminate fitting skills: scribing, patternmaking, undercutting and kerfing.

Each of these essential skills is discussed below in the context of making a site-built countertop (see pp. 36-37 and pp. 40-43). Because the underlayment is already in place, you don't have to worry about fitting a completed top; you can concentrate on fitting the thin and somewhat flexible sheets of laminate. If you are faced with fitting an entire finished shop-built top, you'll find some of these fitting skills useful, though their application may need to be modified slightly.

Scribing

Scribing is the most basic fitting skill, and it also happens to be very easy to do. When you scribe a fit, you merely draw a line on the object you want to fit that is exactly parallel to the surface you are fitting to. If the scribing is done accurately and the cut made precisely, the two pieces will fit together very well.

Laminate is most often scribed to get a close fit against a wall, and to get a consistently parallel line when doing this, I like to slide a small block of wood along the wall with my pencil marking a line tight to the other side of the block (see

To mark a line parallel to the wall, set the laminate sheet in place against the wall, then use a scribe block to guide a pencil. Any scrap of wood can be used as a scribe block; to allow for making a good cut in a laminate sheet, the block should be at least ¼ in. wider than the largest gap between the wall and the sheet.

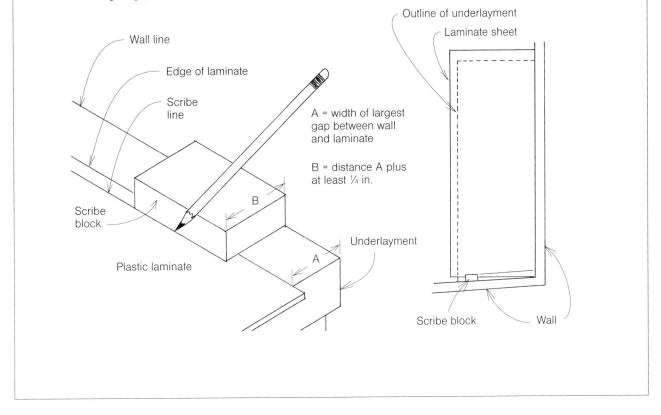

Wall line

Edge of laminate

Scribe line

A = width of largest gap between wall and laminate

B = distance A plus at least ¼ in.

Scribe block

Plastic laminate

B

A

Underlayment

Outline of underlayment

Laminate sheet

Scribe block

Wall

the photo at right). The distance I mark from the wall (and therefore the width of my scribe block) must be at least equal to the widest gap (see the drawing above). If it's a little bigger, that's even better, because cutting laminate close to the edge is difficult; I usually make my block ¼ in. wider than the widest gap. If the laminate is dark-colored and a pencil line would be hard to see, I lay down a strip of masking tape and mark on that. I use laminate snips to cut to the scribed line. For pinpoint scribing to irregular contours, a compass works better than a pencil and block, but the need for complex scribing is rare with countertops, and cutting an intricate profile is very difficult.

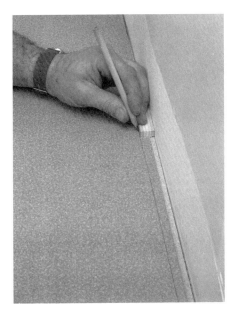

As you slide the block along the wall, scribe a line on the laminate. The line you draw will mirror the contour of the wall.

Scribing a good fit with a countertop section like this one would be impossible. But by using a combination of patternmaking, undercutting and kerfing, you could make a precise fit without a lot of trouble.

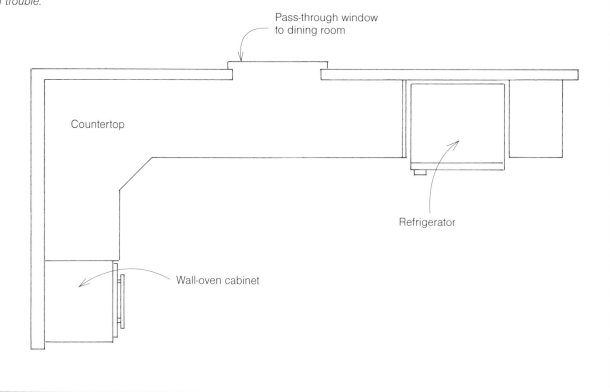

Pass-through window to dining room

Countertop

Refrigerator

Wall-oven cabinet

If there will be an attached ¾-in. thick backsplash at the wall-to-counter juncture, a precise wall fit with the laminate is not critical (within ⅜ in. would be great, though up to ⅝ in. would be acceptable). But a ceramic tile backsplash is thinner, so it's important to get closer (no more than ⅛ in. away from the wall). If you're not satisfied with a scribed cut you've made, you can try it again, but don't cut your sheet short. If your scribed cut is real close but not quite right, you can fine-tune it by marking the "high spots" and block planing or sanding away the little bit of excess.

Patternmaking

Scribing works for most laminate-fitting operations, but it has its limitations. I once resurfaced a countertop in a kitchen with a layout like the one shown in the drawing above. When this counter was first made, the wall-oven cabinet, refrigerator end panel and pass-through were installed and finished off around the already fabricated countertop. For me to fit my laminate sheet into the confines of this configuration required skills beyond scribing because, short of removing the wall oven and the refrigerator side panel, scribing was impossible.

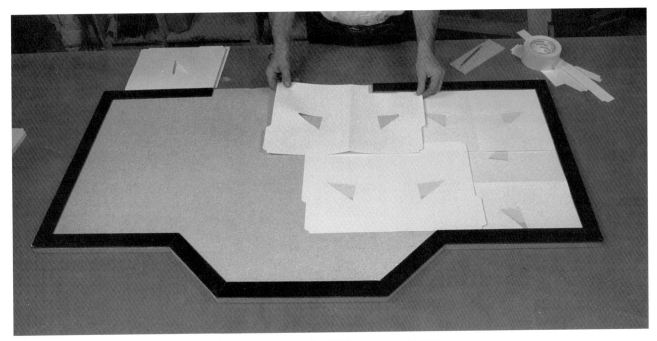

Manila file folders can be assembled into a pattern for fitting a countertop when scribing is not practical, as in this mockup. The triangular cutouts allow each folder to be taped down to the underlayment.

To solve the problem, I borrowed an idea from a vinyl flooring installer I had watched a couple of weeks earlier. He was installing a ¼-in. lauan plywood subfloor in a tight little bathroom and had to fit it around a vanity, tub, and various wall jogs. He could have spent a lot of time measuring and writing down his measurements, transferring his measurements to the lauan, and cutting the sheet to size. (That's what I would have done, and the sheet probably wouldn't have fit as well as I would have liked.) Instead, he cut, coped, fit and taped down pieces of heavy paper, making a full-size pattern that conformed exactly to the shape of the floor. When he was done (and it didn't take long), he simply lifted the pattern off the floor, set it on the plywood, traced the outline, cut the subfloor and set it in place. And it fit the first time. He had made a miracle cut

with minor effort, and I immediately seized upon the laminate applications for full-size patterns.

The photo above shows my method for assembling a pattern on an underlayment mockup with a protrusion for a bay window. For my pattern paper I use manila file folders; they're cheap, stiff and easily worked. Before unfolding each folder, I knife out a small triangle in the center. When I unfold and position the folder on the underlayment, I hold it in place with masking tape (I use 2-in. wide tape) over these cutouts. I cut, fit, tape down and scribe as necessary to make a patchwork pattern of the entire layout. When it's done, I apply masking tape over all the overlapping edges to join the pieces into one big sheet. Then

Taping over all the loose edges of the pattern lets you lift it up in one piece. You can then tape it down over the piece of laminate you need to fit.

I slowly peel up the pattern (see the photo above), lay it on my laminate sheet, pat down the taped triangles to hold it in place, and trace the outline. It takes a little time to put a pattern together, but it's time well spent to get a perfect fit the first time. I now use patterns all the time for many different things and still marvel at the practicality and precision of the approach.

Undercutting

Occasionally you'll need to make a fit that goes beyond the capabilities of a scribe or pattern technique. For example, in the countertop resurfacing job with the pass-through (see the drawing on p. 62), I needed to fit the laminate around the wood casing and jamb that trimmed out the pass-through opening.

To make a precise scribe or pattern fit here would have been difficult at best, but by undercutting the casing and jamb (see the drawing on the facing page), I greatly simplified the cutout profile and allowed myself extra room to work. I made the undercut with a flush-cutting backsaw.

Kerfing

Kerfing is virtually the same as undercutting, but the saw is directed into the material only a short way (see the photo on the facing page). Kerfing is useful when a countertop butts up to a cabinet side panel. A good scribe or pattern fit here is adequate, but a kerf cut into the wood ¼ in. to ⅜ in. (don't go all the way through the cabinet side) is easier to make and will likely look much better.

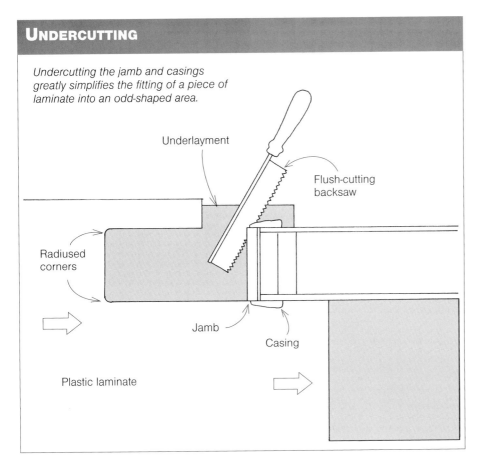

UNDERCUTTING

Undercutting the jamb and casings greatly simplifies the fitting of a piece of laminate into an odd-shaped area.

Underlayment

Flush-cutting backsaw

Radiused corners

Jamb

Casing

Plastic laminate

Also, if the laminate should shrink a bit (it happens), a butted fit will open up, while a kerfed fit will probably not show any ill effects. Any extra width of kerf that may show after glue-down can be touched up with a color putty that matches the wood.

MAKING SEAMS

Another delicate aspect of fitting is making seams. I used to think that making a good seam between adjoining pieces of plastic laminate was the ultimate laminate challenge, but that was because I didn't really know what I was doing. In the early days of my career I would butt two factory edges together and hope it was good enough. If it wasn't, I'd file and sand and fuss over the joint until I

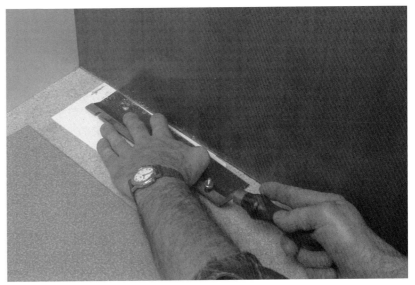

The kerf cut into the side of this tall cabinet will capture the edge of the laminate sheet, resulting in an excellent fit at the side. The flush-cutting backsaw is shimmed up with manila file-folder material to match the thickness of the laminate.

thought it looked acceptable. I had to learn the hard way, but I'm going to save you the trouble.

How can you make a good seam, a seam you can be proud of, a seam that people must search for to see? The secret is a router technique called mirror cutting, which is done with a straight bit. Mirror cutting entails directing the bit down the center of two butted laminate sheets, cutting both edges at the same time. Making this cut in one smooth forward action yields two machined and precisely parallel cut edges. Although you want a mirror cut to be straight, minor variations pose no problem because any slight deviation of the bit into one sheet will be mirrored by a corresponding deviation on the other sheet; the width of the cut will not be altered. Several ways to make a mirror cut are described below. But whatever method you use, the setup work is similar.

First, the two edges to be joined must be relatively straight, and they need to meet each other fairly closely because the cutting bit is not very wide and it needs to cut into both edges. Also, unless you're making a mitered seam, the edges to be joined need to be square to the front edge of the countertop. I square them by marking one laminate edge square with a framing square indexed off the front of the counter, and cutting it with snips. Then I slide the other piece of laminate just under the square-cut edge, and use that edge as a guide to make a line on the sheet below. Then I cut to the line, and I'm ready to seam.

The next order of business is deciding where to make this cut. On a shop-built countertop laminated apart from the cabinets, it's a breeze to make a seaming cut because there is plenty of room to

set up for the cut on top of the underlayment and run the router right on through. However, on a site-built countertop the back wall will be in the way. Sometimes you can pull the underlayment ahead far enough but at his point, it's probably not worth the effort. Instead, when making seams in site-built tops it is usually necessary to move the laminate away from the wall to make the cut. To do this, first position the laminate sheets in place where they will go on the underlayment. The seams should be butted together. Then pencil a registration mark across the joint. When you move the sheets to reposition for seaming, realign the marks to reestablish the proper seaming position.

For site-built countertops requiring miters or other difficult-to-position seams, make the seam through the rough-cut laminate pieces first, apply tape to hold the edges tightly together, and then position a full-size underlayment pattern over the laminate. Trace, cut, check the fit, and you're ready for glue-up.

Mirror cuts may be made by cutting along a straightedge, with a seaming jig or with a seaming base; you can also make seam cuts using an underscribe trimmer base. With all these methods, a router and bit are directed through two butted pieces of laminate, but the methods differ in how easily the cut is set up and how the cutter is guided through the laminate.

Mirror cutting along a clamped straightedge

You can make a decent mirror cut by guiding a router equipped with a ¼-in. straight bit along a clamped straightedge (see the photo on the facing page). Slide a piece of thin plywood under the cutline and adjust the cutting depth so the bit just slices through the laminate.

This arrangement requires no special equipment, and I've used it to make many mirror cuts over the years. However, setting up for the cut does require a lot of time.

Mirror cutting with a seaming jig

A much quicker and easier way to make mirror cuts is with a seaming jig. Seaming jigs can be purchased (see Resources on pp. 127-129) or you can make your own. I thought about making a jig for a long time but wasn't quite sure how to put it all together until I noticed something called T-tracks in a Woodsmith Shop catalog (see Resources on pp. 127-129). These aluminum channel pieces are sold specifically for jig applications and they were exactly what I needed to make adjustable laminate-sheet hold-downs for the jig of my dreams. I immediately ordered two

Mirror cutting along a clamped straightedge does not require specialized equipment, though it involves a lot of setup time. *Photo by Kevin Ireton.*

The author built this seaming jig out of aluminum T-track, four knob-and-washer hold-downs, and scraps of wood. Total cost for the hardware was about $50.

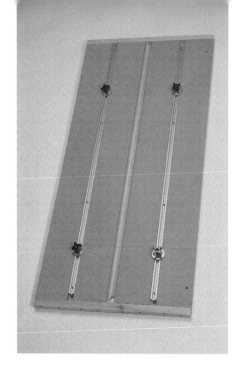

36-in. long T-tracks, and when they came in, it took me less than an hour to put together the jig, which is shown in the photo at left.

I made my seaming jig by sandwiching the ¾-in. high T-tracks between pieces of ¾-in. MDF. A base of ¼-in. plywood and pine end cleats support the assembly (see the drawing below). Screws and glue hold the components together. T-tracks have mounting holes in their bottoms, but the ¼-in. plywood base doesn't provide adequate purchase for screws, so I glued the tracks in with construction adhesive, and the jig has held together well.

SEAMING JIG

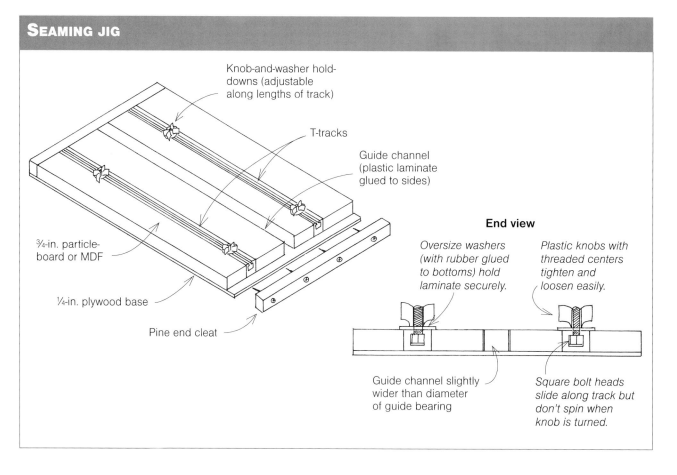

Knob-and-washer hold-downs (adjustable along lengths of track)

T-tracks

Guide channel (plastic laminate glued to sides)

¾-in. particle-board or MDF

¼-in. plywood base

Pine end cleat

End view

Oversize washers (with rubber glued to bottoms) hold laminate securely.

Plastic knobs with threaded centers tighten and loosen easily.

Guide channel slightly wider than diameter of guide bearing

Square bolt heads slide along track but don't spin when knob is turned.

The center guide channel needs to be wide enough to accommodate the guide bearing on a trimmer bit; I made mine $7/8$ in., which is wide enough for most bits. Before assembling the unit, I glued laminate to the MDF channel edges to provide a smooth and durable surface for the roller bearing to roll along.

Plastic hold-down knobs (also available from Woodsmith) have a threaded center to accommodate standard $1/4$-in. - 20x $1 1/4$-in. hex-headed bolts. T-tracks are essentially a channel that the square-faceted bolt heads can slide along, but not spin around in or fall out of. Screwing the knobs down tight exerts

clamping pressure on the laminate by means of oversize washers (available at any hardware store). I glued a piece of rubber to the bottom of the metal washers so they grip tightly, but don't do any damage. The hardware for this jig cost me around $50.

My seaming jig has proven itself to be an invaluable tool. Now, when I want to make a seam, I lay the sheets over the jig with their edges butted together over the center cutting section, adjust the integral hold-downs to hold the laminate in place securely, and zip down the joint with a router and bearing-guided bit (see the photo below). When making

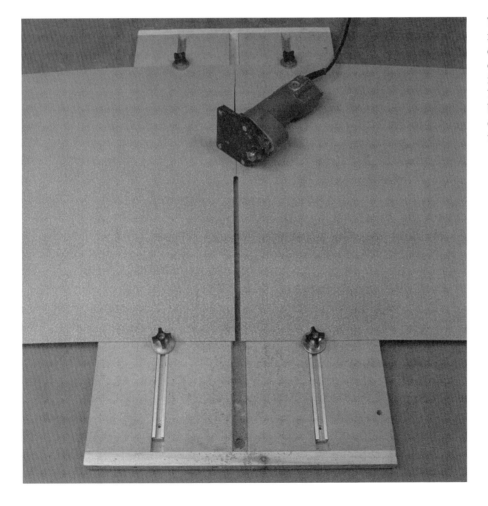

With the edges of the laminate sheets butted together and clamped over the center of the seaming jig, the router is passed along the seam. The bearing on the router bit is guided along one edge of the channel underneath the seam.

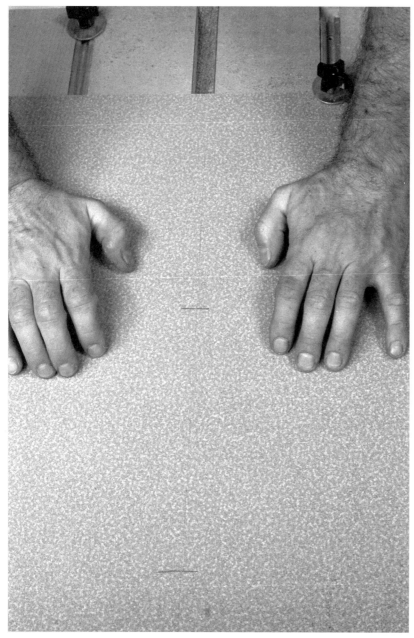

After making the seaming cut, butt the pieces together and slide them back and forth against each other to find the best fit. Then make registration marks across the seam to facilitate repositioning when you glue the pieces down.

a cut with a seaming jig, make sure to direct the router through the cut with the bearing held firmly to one side— your left side when pushing the router through, or your right side if pulling it toward you. Once the seam is cut, fit the pieces together (see the photo at left), then make registration marks across the seam so you can realign the pieces when you glue them down on the underlayment. It's that easy.

Mirror cutting with a seaming base

Another mirror-cutting option is to use a laminate trimmer with a seaming base (see the photos on the facing page). To use a seaming base, two adjoining sheets of laminate are clamped down to a flat surface, and one sheet is positioned to overlap the other by 1/8 in. to 7/8 in. The seaming base has a guide lip integrated onto the bottom that hooks over and rides against the edge of the lower sheet as the tool is moved down the center. A 1/8-in. router bit machines a mirror cut down the center of both sheets. Since the cut is made with one sheet overlapping, the cut on that sheet actually has a small back bevel (the same undercut could also be achieved in the seaming jig by overlapping the sheets). Because one laminate sheet edge acts as the straightedge guide for the tool, it must be relatively straight but not perfect because it also gets cut in the process.

Seaming with an underscribe trimmer base

A seaming base should not be confused with an underscribe trimmer base. Although the tools look almost identical and are set up and used virtually the same way, an underscribe does not make a mirror cut. Instead, the integrated guide lip on the base follows the

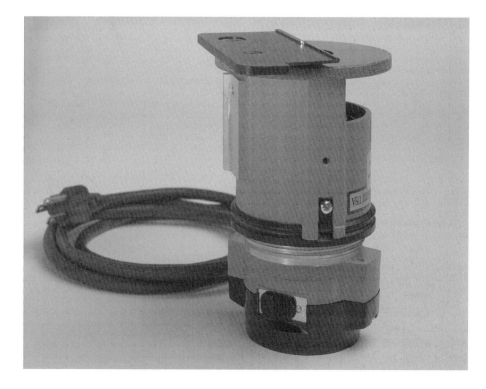

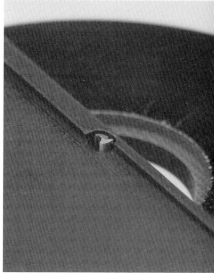

The Betterley seaming base
(left) uses a ⅛-in. cutting
bit and an offset guide lip
(above) to mirror-cut
two overlapping sheets
of laminate.

lower sheet of laminate and cuts only
the top overlapping sheet. If the guide
edge on the lower sheet is not perfectly straight, the seam will not be
perfect either.

Underscribe trimmers are designed to be
used after the laminate is glued down;
the cut edge is made to lie precisely next
to the straight edge after the trimmer is
run through. In theory, no chips from
the cutting process will blow under the
preglued overlap section, but in practice
tests, I've found that some dust does.

A bigger problem than dust is the fuzzy
remainder of phenolic material that the
bit peels back and leaves hanging on the
underside of the laminate. It's enough to
prevent the just-sliced piece from sticking down level with the adjoining piece
of laminate. This extra material should
be lightly sanded, filed or scraped off

before you glue down the sheets. Make
sure not to remove any edge material
other than on the bottom corner when
doing this, and make sure that the
scrapings don't land in the glued
surface below.

To make the best possible seams with
the underscribe trimmer, make two cuts
in the unglued sheets: the first pass
using a factory straightedge or the
straightest edge you can make as a
guide, and the second pass in the other
direction using the just-made edge as
a guide.

GLUING AND TRIMMING LAMINATE

Having cut the laminate pieces to fit and made any necessary seams, you are now ready to glue on the laminate, trim off the overhang and finish-file the edges. This is the part of the job where any mistakes will usually show as painfully obvious flaws on the finished counter. But if you plan your work well, take your time and follow my instructions, you will meet with success.

PLANNING THE ORDER OF WORK

A countertop section, consisting of several edges and a top, is not glued up all at once; it is a step-by-step process that should be taken in logical order. Begin with your edge pieces—glue them on and trim off the overhang before putting on the top pieces. If you apply the top piece last, there won't be an exposed glue line on the top surface and thus no way for liquid spills to penetrate the assembly (see the drawing below left). Some fabricators reverse the order and apply the edges after the top piece, but I question the wisdom of that arrangement.

Where two edges meet at an outside corner, one laminate edge must be glued, bonded, rolled down (I'll explain these procedures shortly) and trimmed before the same is done to the adjoining side, because one edge piece must overlap the end of the other. Side pieces typically go on first, so unsightly phenolic end grain won't be visible from the front (see the top drawing on the facing page). On edge pieces that meet at inside corners, it is not as necessary, but it's easier to apply and mill the top and bottom overhang of one edge before applying the other.

Because there are usually several pieces to deal with, countertop edges are time-consuming to install. To keep the job

EDGE STRIPS BEFORE TOP SHEET?

Countertop edge strips can be applied before or after the top sheet of laminate. Installing the edges first is better because no glue lines are left on the top surface, where they are vulnerable to moisture intrusion.

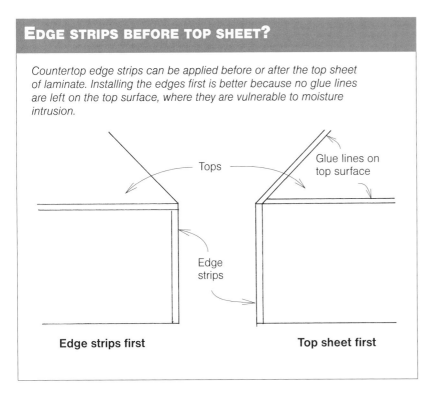

Tops

Glue lines on top surface

Edge strips

Edge strips first

Top sheet first

rolling along, proceed in a logical sequence, and work on more than one edge at a time—glue in one area, and while the glue sets, trim in another. The drawing at bottom right shows an example of how to go about planning your work.

With plastic laminate, cleanliness is critical to successful glue-up. Before you start, make sure the surfaces to be glued are free of imperfections and dust. On edges I typically make a couple of cleaning swipes with a sharp block plane or a sanding block wrapped in 36-grit paper. On the deck surfaces and sheets of laminate, I always do a final wipe-down by hand; fingertips will detect imperfections the eye cannot readily see and remove residual dust a vacuum cleaner or broom sometimes leaves behind.

As you're gluing in one area while trimming in another, watch out that you don't blow dust and chips onto nearby glued edges or pieces of laminate. If this does happen, wait for the glue to dry completely and brush off the grit. If there is a lot of particulate contamination, you may want to recoat the pieces after brushing them off.

Although gluing and trimming are done concurrently during the fabrication process, we will deal with them individually here.

GLUING WITH CONTACT ADHESIVE

Plastic laminate is applied using contact cement. If you read Chapter 1, you already know how contact adhesives work, that they are available in solvent-based and water-based forms, and that they can be applied either by spraying, or manually with a brush and roller. My emphasis here will be on the manual approach because it is a simple, low-cost and effective method used by beginners and professionals alike.

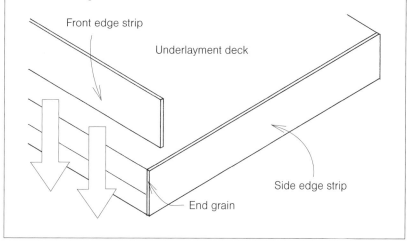

EDGE STRIPS AT CORNERS

When two edge pieces meet at an outside corner, one piece is glued on and trimmed flush before the other is applied. Side pieces typically go on first, so the front edge will cover the dark phenolic end grain of the side.

Front edge strip

Underlayment deck

Side edge strip

End grain

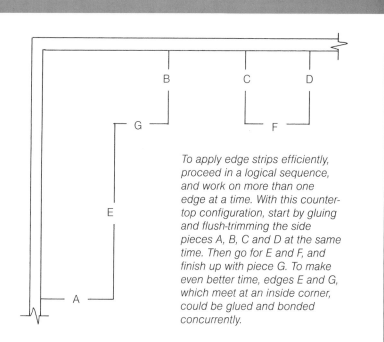

EDGE-STRIP INSTALLATION SEQUENCE

B   C   D

G   F

E

A

To apply edge strips efficiently, proceed in a logical sequence, and work on more than one edge at a time. With this counter-top configuration, start by gluing and flush-trimming the side pieces A, B, C and D at the same time. Then go for E and F, and finish up with piece G. To make even better time, edges E and G, which meet at an inside corner, could be glued and bonded concurrently.

The first rule of gluing is to read the instructions on the container of adhesive you're using. Though all contact adhesives work similarly, there are some differences among formulations, and you will want to do everything you can to get the strongest bond from your particular glue.

A word about aerosols

Before discussing manual application, I want to inject here some important advice concerning aerosols, because you will undoubtedly want to try them on occasion. The general rule of thumb when spraying any contact adhesive is to shoot for 80% coverage, but one adhesive company representative confided to me that it's much better to have closer to 100% coverage. A cynic would say that's because he wants to sell more glue, but I don't think that's the case in this instance. The problem with aiming for 80% coverage is that there's no way to know you haven't sprayed on 70%, or even 65%. And even if you do manage to get an adequate 80% on, in places where moisture penetration is a possibility (i.e. surface seams and edge seams), 80% coverage means that 20% of the underlayment under the seam does not have that little extra barrier, which could make a big difference.

My advice, and my standard procedure, when spraying a contact adhesive is to follow the directions on the can, but if the label says 80% coverage is adequate, I disregard that and shoot for 90% to 100% on the wide open spaces, and a for-sure 100% coating along edges and surface seams. (With brush and roller application, the issue doesn't arise; 100% coverage is a given.)

Brush and roller application

Regardless of the kind of adhesive used, it is important to apply an even layer. With that thought in mind, I have found there is a decided difference between manually applying solvent- and water-based adhesives—water-based adhesive has a thinner consistency and goes on smoother and easier than solvent-based adhesive.

Selecting your equipment Solvent-based adhesives and water-based adhesives require different applicators (see the photo at top right on the facing page), and using the wrong type for the glue can make your work a lot harder. One guy I used to work for would issue me 3-in. wide disposable foam brushes to apply solvent-based adhesive. After a couple minutes of use, the foam would become limp and start to dissolve, and the glue layer was far from smooth. On the other hand, a foam brush works great for water-based adhesive; I wouldn't use it to cover large areas, but on edges, a 2-in. wide disposable foam brush (see the photo at bottom left on the facing page) will get you through one job without any problem. If you won't be getting to all the edges within a few minutes, you can wrap the brush in plastic wrap to keep it from drying prematurely. Bristle brushes (natural bristles are a must with solvent adhesives) will work just as well but they're more expensive. And it's not worth trying to clean brushes used with any contact adhesive.

For gluing up the decking and sheets of laminate, a brush will work, but I recommend a paint roller because it does a better job on large surfaces. For rolling water-based adhesives I use inexpensive short-nap ($^3/_{16}$-in.) paint roller covers. For solvent-based adhesives, it's best to use more expensive stippled-nap roller covers made specifically for rolling out the heavier bodied liquid. Such rollers also better withstand the destructive action of the solvents.

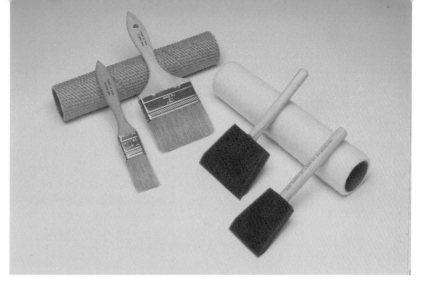

As with the foam brush, you can wrap roller covers used with water-based adhesive in plastic to prolong their usable lifespan. Some fabricators leave their brushes and rollers submerged in the can of adhesive (a five-gal. container for a roller) when not in use and can get a lot of mileage out of them.

To save the time and trouble of using a paint-roller tray, I usually pour some adhesive directly on the surface being glued, and roll it out. A plastic squeeze bottle (see the photo below right) is a handy dispenser for water-based adhesive, but not for solvent-based adhesive, which is too viscous to flow through the

A regular paint-roller cover and common foam brushes (right) are the best tools for applying water-based adhesives. For applying solvent-based adhesives, a special stippled-nap roller with a tough, pebbly texture and natural bristle brushes (left) work best.

A foam brush is an inexpensive and adequate tool for applying water-based contact adhesives to counter-top edges.

A plastic squeeze bottle makes a handy dispenser for water-based adhesive that is to be spread out with a roller.

nozzle. I started using squeeze bottles after accidentally spilling a whole gallon of adhesive on a kitchen floor.

Applying contact cement You will be applying glue to both the laminate and the underlayment, then allowing the adhesive to dry before making contact between the two. It is often easiest to use the underlayment as a work surface for spreading glue on the laminate. When this is done, the sheet can be set aside to dry where it won't be disturbed, and you can focus on spreading glue on the underlayment.

When manually spreading contact cement, the layer of glue, also called the glue line, need not be thick, but it must be "sufficient." How much is sufficient? Adhesive manufacturers recommend that their adhesive be applied so that a given amount (typically 2 grams to 3 grams per sq. ft.) of adhesive solids remains after the solvent has evaporated. When I asked one adhesive-company representative how anyone would know if the right amount had been applied, this is what he told me: Take a sample square-foot piece of underlayment, weigh it, apply a representative amount of adhesive, and then weigh the sample again when dry. The difference is the weight of the adhesive solids. His advice was reasonable, but not very practical or useful for the average fabricator—not many of us own a set of laboratory scales. It turns out that the recommended coverage is a minimum geared to big-time cost-conscious fabricators who don't want to use a speck more adhesive than is necessary (it's probably the technical equivalent of 80% coverage).

For small-scale fabricators who apply adhesive with a brush and roller, getting adequate coverage is rarely a problem. In fact, we probably use a tad more adhesive than is really needed, and that's perfectly okay. The important thing is to apply an even layer.

As a rule, porous surfaces, like particle-board edge grain, should always have a couple of coats of adhesive, but one adequate layer on the deck and laminate is fine. A roller will typically put down an adequate layer of adhesive if you roll it on, and roll it out smooth and even. Don't roll it on thick, but don't roll it on as thin as possible, either. If in doubt about whether or not sufficient adhesive has been applied, put on a second thin coat when the first has dried (see the section below). While it is possible to put on too much adhesive, if you stick to these recommendations, you shouldn't have a problem.

Never apply contact adhesive in direct sunlight because the heat will cause the surface to dry too quickly, trapping un-cured adhesive below the dried layer and resulting in a weakened bond. Flashing off, as this phenomenon is called, is of particular concern with solvent-based adhesives, and it can also occur if the adhesive is applied too thickly.

Drying times

Assuming that the glue line is uniform and sufficient, the next trick to getting a strong bond is to make sure the glue is dry, but not too dry, before making contact. If the glue is not dry enough, solvent will be trapped between the mating pieces and interfere with a good bond. If the glue is permitted to dry too long before contact is made, the bond will also be weak.

All water-based adhesives undergo a color change as they dry, and I consider this another advantage to the product. 3M's Fastbond 30-NF comes in either a white or blue-green color. Water-based adhesives that are white in the can, such as Lokweld's H_2O brand, will dry clear, and the blue-green dries to a distinctly different shade of green.

Solvent-based adhesives do not change color, though there is a perceptible difference in the sheen of wet and dry glue. The best way to tell when a solvent-based adhesive glue line is dry is to touch it with your knuckles or a piece of paper (don't use your fingertips, since they are more likely to have natural oils that will interfere with the test). If the glue is tacky but doesn't transfer to your skin, it's ready to bond.

Another test for dryness is to sight across the glued surface; with either kind of adhesive there should be a uniform and slightly glossy sheen. If the surface is splotchy looking, some areas of adhesive may have soaked in more than others, and an additional, thin coating may be called for.

The critical time frame for bonding once the adhesive has dried is often referred to as the open time, or working time. Open time varies among brands of adhesive, and this information can be found with the instructions on the container. In general, most contact adhesives have one hour of optimum contact ability once the glue has dried.

The most important thing to know about open time is that the strongest possible bond is achieved when the adhesive has just finished drying, and potential bond strength decreases the longer the glue is allowed to dry after that. If for some reason, you neglected to make adhesive contact within the recommended open time, it's not a problem because the pieces can be recoated with adhesive (again, be sure to use only a thin coating).

Applying the edge strips

Positioning and joining laminate pieces to the underlayment can be a delicate operation because of the touch-and-grab nature of contact cement. As discussed on p. 72, the edge strips are applied first, then the top piece.

A long glue-coated edge strip can be applied by one person. Hold the strip loosely coiled in one hand, while positioning it along the edge with the other.

To apply edge strips, hold them parallel to the underlayment edge, touch one end down, then progress along the edge, making contact in approximately 1-ft. increments. Make sure to maintain sufficient laminate overlap (which will be trimmed off later) at the top and bottom. Edge strips that abut walls or inside corners should be positioned in those areas first, then worked along from there. Applying long lengths of laminate edging is easier if you have a helper, but one person working alone can manage an unwieldy strip of edging by rolling it into a loose coil and holding it in one hand while laying it in place with the other (see the photo above).

Some fabricators apply edge strips flush with the bottom of the underlayment so they have to trim only the tops. There is nothing wrong with using this approach as long as your strips have a smooth and straight bottom edge and you can consistently maintain flush bottom placement as you lay the strip into place.

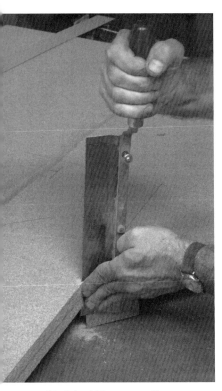

At this 45° corner, the corner strip of laminate is applied first. A kerf at each end (cut with a flush-cutting backsaw) helps in fitting. The laminate edges that meet the angled corner will have 45° bevels cut on their ends.

I prefer to overlap the top and bottom because it demands less precision, and I feel a cleanly routed bottom edge looks best. However, if it is logistically difficult to rout off the bottom overhang, I too match up the bottom edges.

When an edge strip fits between two inside corners, it's helpful to cut a shallow kerf at both ends (see the photo at left). This allows the strip to be cut slightly oversize, and ensures a good fit.

Edge strips at radiused corners

Applying edge strips of laminate to radiused corners may require some special techniques because the laminate must bend around the curve. Horizontal-grade laminate will usually bend to a radius of 9 in. without any problem. Post-forming grade laminate will, because it's not as thick, bend to a smaller radius at room temperature. To make tighter curves with either laminate, you can direct a heat gun onto the piece as you're laying it in place around the curved edge. However, although the horizontal-grade laminate will take a little tighter curve with heat, it won't bend nearly as far as the post-forming grade. If you want to heat-bend a piece of horizontal-grade laminate, experiment to determine the material's practical limits. For really tight curves, a strip of post-forming laminate should easily bend down to a ½-in. radius with proper heat. See the sidebar at right for a way to prebend tight curves.

Applying the top sheets

Top pieces are larger than edge strips and therefore more difficult to lay in place, so it's best to distribute several temporary spacers over the countertop underlayment, set the laminate on top of the spacers, shift it around into precise position, and then pull the spacers out.

POST-FORMING A TIGHT RADIUS

Post-forming is the process of bending a piece of laminate around a form using heat and pressure. Although it isn't possible for the small-scale laminate fabricator to heat-bend large sheets of laminate, edge strips can be successfully post-formed to conform to a very small radius without any specialized equipment. A ¾-in. radiused curve is ideal for this application, and I have found I can get a better edge fit by pre-bending a ¾-in. radius using a household iron and a form (see the drawing at right). As shown in the photos at right, I set the iron on high and place it directly on the face side of the laminate for 30 seconds, then I bend it by hand around the form (you might want to wear gloves because the laminate is very hot). Use a caul board and spring clamps to distribute pressure and hold the piece in place for a few minutes until it cools. Once the curve is formed, it's a simple matter to glue the laminate strip in place.

I recommend you experiment with your own iron on a strip of laminate to find the optimal heating time. The laminate should bend around the form without any hairline cracks forming— these indicate that the piece was overheated. Once a section is heat-bent, it should not be heated and bent again. And, of course, remember that only a post-forming grade of laminate should be used for post-forming.

Post-forming radius jig

The jig is about 12 in. by 10 in. and made of a double layer of ¾-in. particleboard. Make the corner radius first, then modify the jig as noted.

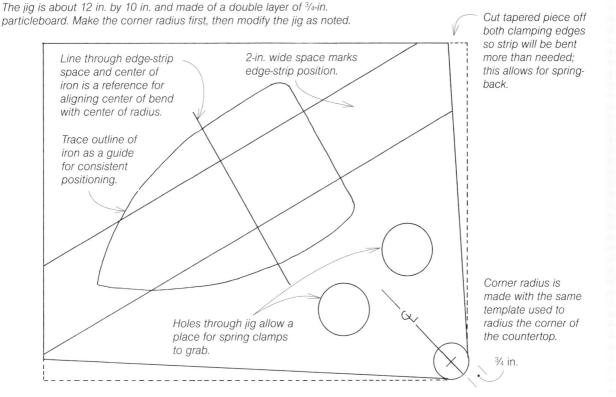

Line through edge-strip space and center of iron is a reference for aligning center of bend with center of radius.

Trace outline of iron as a guide for consistent positioning.

2-in. wide space marks edge-strip position.

Cut tapered piece off both clamping edges so strip will be bent more than needed; this allows for spring-back.

Holes through jig allow a place for spring clamps to grab.

Corner radius is made with the same template used to radius the corner of the countertop.

¾ in.

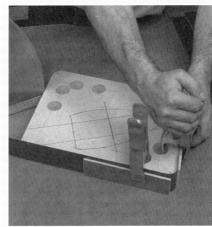

Hold a hot iron directly on a strip of post-forming laminate for about 30 seconds (far left), then bend the heated laminate around the ¾-in. radiused form and clamp it in place until cool (near left). The small strip of ¼-in. plywood acts as a caul and distributes the clamping pressure.

Venetian-blind slats make great temporary spacers. Spaced 6 in. to 10 in. apart, they allow larger sheets of laminate to be placed on the glued underlayment and shifted into position without sticking.

I use metal venetian-blind slats for temporary spacers (see the photo above) because they're durable and can be assembled into a compact bundle for storage. (They're also free if you can find an old discarded venetian blind, but those are becoming rare.) Plastic slats from modern mini-blinds can also be used, although they don't give as much lift to the laminate sheet, and you'll need to space them closer together. In lieu of slats, you could use dowels, thin strips of wood or stiff cardboard, or whatever you can find that will hold the surfaces apart, is clean, and can be tugged on without coming apart.

When pulling the spacers out, start at one end and work to the other (see the photo on the facing page), or in the middle and work out in both directions, pressing down the laminate as you go. Don't start at the ends and work toward

the middle, or you may entrap air pockets. By the time you got to the middle, the last bit of laminate might be humped up, and pent-up internal stresses would stubbornly resist your desire to make it lie flat. Although you might be able to force it down, it would probably pop back up eventually. I've been down that road.

When using low spacers like the slats, make sure to pull all the spacers out. I once worked with a fellow who inadvertently "lost" a slat under his laminate at an inside corner. After the laminate was all stuck down, he discovered a mysterious long lump under the surface.

Applying seamed pieces
To glue up and apply sheets that meet at a surface seam (and this includes seamed edge and backsplash strips too) I've found it is quicker and better to glue up the mating pieces at the same time.

After the laminate is in position, pull the spacers out one by one, starting at one end and working to the other. Smooth the laminate down by hand as you go.

Make sure not to let adhesive get on the joining edges because it will hold the sheets apart enough to ruin the perfectly milled fit you made (see pp. 65-71). Any glue that does inadvertently make it onto the edges can be rubbed off with your fingers or a rag dampened with adhesive solvent.

With the help of some spacer slats (see the discussion on the facing page), position and glue down the larger of the two sheets; if one sheet is fitted into a corner or other tight space, do that sheet first. Then position the second sheet over the slats, align the reference marks you made on the sheets when you cut them (see the photo on p. 70), press down firmly at the seam, and proceed to lay the rest of the sheet in place.

If, after making contact at the seam, you find the fit isn't as tight as you know it can be, here's a little trick that may help.

Often you can provide a little push to tighten the joint by skipping over a slat or two from the seam, bonding down the laminate for about 1 ft., then pulling out the slats you skipped and applying pressure to the slightly humped-up laminate by pushing it down and toward the seam. The seam should tighten up nicely. This technique, however, is not without risk; the slight hump formed by the slipped-out spacers must indeed be slight. If it isn't, you're in bigger trouble than before.

Applying pressure to the glue line

Contact adhesive sticks on contact, and a hand pat will hold the laminate in place; but to ensure a permanent bond, much greater pressure must be applied. Large production fabricators run their tops through a fairly expensive piece of machinery called a pinch roller, which applies a lot of pressure uniformly over every square inch of the countertop. It's an advantage they have over us small custom fabricators, but that doesn't mean we can't do a good job. We just have to work a bit harder at it.

The best way for a small fabricator to apply the pressure needed to knit the two adhesive surfaces together is with a rubber hand roller. Such tools, made specifically for rolling laminate, are usually available from laminate suppliers in a variety of styles and sizes. A 3-in. wide roller is considered to be the most efficient. There are wider rollers, but they will not give the concentrated pressure that a 3-in. roller with maximum body pressure behind it will deliver. The Gundlach V300 pressure roller (see the photo at left) is widely regarded as the best two-handed roller for applying maximum pressure over large surfaces, especially when reaching. The Sabel roller (see the photo below) is a compact and comfortable-to-use one-handed tool that works great on edges as well as open surfaces, although it may not be as effective when reaching ahead to roll the back half of the countertop.

Rolling laminate is hard work. Pressure need not be sustained, but every square inch must be rolled, and to exert maximum pressure you've got to put your weight into it. I roll the whole top and edges twice, once before routing off the overhang, and again after routing. I make a concerted effort on the top edges and at seams. It's easy to skip this step because the laminate appears to stick fine with a simple hand pat, but don't be deceived. Many adhesive fail-

The Gundlach V300 pressure roller has an offset handle bend that allows the fabricator to get a good grip for applying maximum pressure when rolling the glued-down laminate.

The Sabel roller is ideal for doing edges, and it works well on tops too.

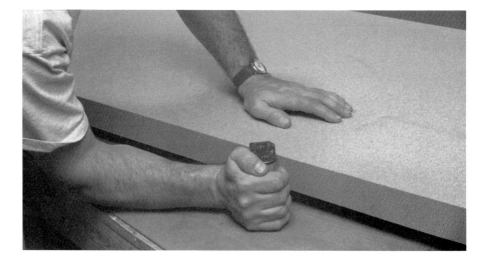

ures, especially with water-based adhesives, can be attributed to a neglect for proper rolling during the fabrication process. So put the pressure to it.

Another way to apply pressure is with a deadblow mallet (see the photo at right), or with a hammer and a block of wood, but this method is not nearly as effective as rolling. To illustrate this point, a Wilsonart technical bulletin recommends that the reader take a sheet of bubble wrap, lay it on a flat surface and try to pop the bubbles with a hammer and block. Then take a 3-in. wide rubber roller to a section of bubbles. I tried it, and the roller is far and away the better tool for the job.

After rolling the laminate, you may be curious to know how well the glue is holding, so you lift up a bit on the edge. Don't be disappointed if the bond isn't as strong as you think it should be. As long as the sheet is stuck down flat, the glue will cure and become stronger over the next day or so. Some adhesives have a higher initial tack than others, and this is a reassuring attribute, but cured bond strengths pretty much equal out over the long run.

WHEN DISASTER STRIKES

In a perfect world, if you followed every instruction on the adhesive can and worked with impeccable technique, all your glue bonds would be incredibly strong to the point of perfection. But life isn't like that. In the real world, through your own errors or circumstances completely beyond your control, you inevitably will one day experience some sort of adhesive bond failure. When it happens, you may know exactly what you did wrong and will learn from the experience. Or you may have no idea why things went wrong—you'll probably conclude that "the damn glue was no good." Either way, the most pressing question will be how to solve the prob-

A deadblow mallet can be used to apply pressure to the glue bond, but it is less effective than a roller, especially on large top surfaces. With any other hammer, use a wood block to protect the laminate from damage.

lem. Fortunately, adhesive bond failures are fixable, either by regluing or by removing and replacing the sheet. I know—I've had a few glue failures in my day.

Regluing a loose sheet

Once my helper accidentally knocked over an open quart container of mineral spirits as we were gluing the edge strips on a kitchen countertop, and a substantial amount of the solvent soaked into the particleboard underlayment. We mopped up what we could, and in a half-hour the spot looked and felt dry enough, so we didn't give it another thought and glued down the top sheet. A week later the customer called to let me know there was a large loose section of countertop. I was mystified as to why, but when I checked it out, I found it was right where the thinner had spilled, and I realized that entrapped solvent had compromised the bond.

The first thing I did to fix the spot (and the first thing you should try if, for whatever reason, you get a loose piece of laminate) was to heat the spot with a household iron and reroll the area; heat will reactivate the glue line. Adjust the

Contact-adhesive solvents can be divided into two categories: Bond breakers, which quickly dissolve contact cement and remove stuck-down pieces of laminate, and cleaners, which are used to remove excess dried adhesive from countertop edges after trimming. Although bond-breaker solvents can be used as cleaners, don't try to use cleaning solvents as bond-breakers.

BOND BREAKERS

If you ever need to lift a piece of glued-down laminate, your best bet may be to use the solvent recommended by the manufacturer of the adhesive (this information is sometimes found on the adhesive container). if you don't have the recommended solvent or if no solvent is recommended, acetone will probably do the trick. I use acetone for various things and

like it because it doesn't smell too bad to me and it evaporates very quickly and completely. However, acetone will also dissolve cabinet finishes and vinyl flooring, so it should be handled with care.

Lacquer thinner will usually work as a bond-breaker too, but it is also a threat to some finishes, and I find the vapors particularly objectionable. The main disadvantage to lacquer thinner is that it does not evaporate as quickly as acetone, so if you plan to reglue surfaces that have been bond-broken with lacquer thinner, you will need to wait until they are fully dry (about 30 minutes) before regluing.

CLEANING SOLVENTS

For a cleaning solvent, I like to use mineral spirits. Paint thinner, as it's also known, is a "safe" ad-

hesive solvent in that it doesn't appear to harm cured finishes or other building materials. But mineral spirits should not be used for bond-breaking purposes because it only softens the glue bond rather than dissolving it, and it is very slow to evaporate.

Another cleaning solvent you might want to try is 3M's Citrus-Base Industrial Cleaner. This orange-scented chemical solution comes in a handy aerosol can, so you won't spill the contents if you accidentally knock the can over. Although the citrus base lends a pleasant smell to the product, avoid inhaling any more of it than you have to, because, like all these solvents, it has an unhealthy VOC content. You should also be aware that all of these solvents are flammable, so take appropriate precautions.

heat to a medium setting (which is typically the setting used for polyester and cotton blends—around 200°F to 250°F) and run it over the poorly bonded spot. It's been my experience that the plastic laminate will withstand this heat without scorching or blistering, even if the iron is left in place for several minutes. But to be safe, I move the iron about in slow motion and put a cloth or paper towel under it. When the area is quite warm, set the iron aside and roll the laminate until it is cool. If the glue line was sufficient to begin with, this repair will work like a charm, and provide a permanent fix. Unfortunately, with my particular problem, the integrity of the glue in that area was almost nil, and the iron-down repair lasted about long enough for me to drive out the driveway. So I had to remove the sheet and reglue it.

Removing a loose sheet

Although contact cement is a downright tenacious adhesive, it will cleave with surprising ease when a bond-breaking glue solvent (i.e., acetone) is introduced into the glue line (see the sidebar above).

To lift a sheet that is completely glued down, carefully work the edge of a thin-bladed putty knife just under the laminate and drizzle some acetone onto the top of the blade so that it's directed into the glue line. The acetone works quickly, and if you continue this process along a section of the edge, working the blade around and farther in as the solvent does its job, before long you'll be able to get a fingerhold on the sheet and lift it up a bit. At this point, continue to feed the solvent into the receding edge of the stuck-together glue line. I have found a spray bottle works great for dispensing the acetone. If you lift up the sheet only slightly and let the bond-breaking solvent to do all the work, the sheet will be free in no time. It's then a matter of recoating and rebonding the surfaces. With my spilled-thinner snafu, I lifted the top sheet of laminate just enough to get to the bad spot, directed a heat gun onto the area to help evaporate any of the spilled thinner solvent that may have still been there, and reglued. The fix has now held for several years.

The only problem with acetone is that it destroys many plastics; before long, the spray-bottle mechanism will be ruined. (I've never known acetone to damage plastic laminate, however.) Although it's possible to get an acetone-resistant plastic bottle, I've yet to find a spray head that will hold up to the solvent. But the cost of a spray head is a small price to pay for the service rendered. I have heard of people using squirt-type metal oil cans to dispense bond-breaking solvents, but I have never used one myself.

TOOLS FOR TRIMMING LAMINATE

A router is the tool best suited for trimming the overhanging edges of glued-down pieces of plastic laminate—in fact, it's your only choice. You don't need a top-of-the-line model, though; any hand-held router that will accept the 1/4-in. dia. shank of a laminate trimming bit will do.

When I was starting out in the building trades, my wife bought me my first router, a 7/8-hp Craftsman from Sears, as a birthday present. It was an inexpensive but sentimentally valuable tool that I learned with and used to make many things, including several countertops. However, that router was designed as a general-purpose tool for a wide range of woodworking operations, and as I

began to do more laminate work, I realized that I really needed another router, one better suited to the peculiar demands of a laminate fabricator. I needed a laminate trimmer.

Laminate trimmers (see the drawing below) are a bit shorter and much smaller around than the average router, so you can operate them comfortably with one hand. A laminate trimmer also has a proportionately smaller base, which makes it considerably more convenient for fitting into small spaces or trimming close to fixed objects, such as walls. More often than not, the base is square to allow for easier and more accurate alignment when guiding the trimmer along a straightedge, as when making mirror-cut seams.

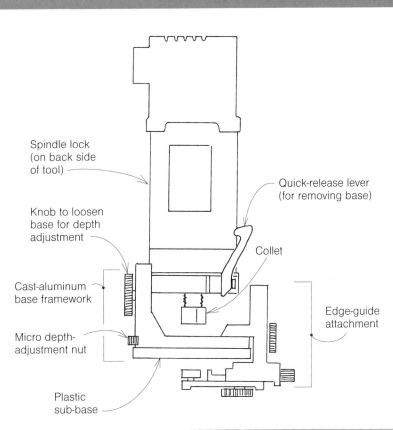

ANATOMY OF A LAMINATE TRIMMER

Spindle lock (on back side of tool)

Quick-release lever (for removing base)

Knob to loosen base for depth adjustment

Collet

Cast-aluminum base framework

Micro depth-adjustment nut

Edge-guide attachment

Plastic sub-base

In 1994 I had the rare opportunity to use, review and evaluate all the name-brand laminate trimmers that were then on the market. My findings were published in an article I wrote in *Fine Home-building* magazine (issue #95). I found certain features particularly endearing. If you're in the market for a laminate trimmer, here are some points to keep in mind when comparing different brands.

POWER

Motor amperage is a fairly reliable indication of power levels, and the laminate trimmers I tested range from 2.8 amps to 5.6 amps. Obviously, the more power, the better, especially if you intend to use the trimmer for other more demanding work, such as routing a molded detail along the edge of a board. But you don't necessarily need all that power to trim laminate. One trimmer I own is rated at 4 amps and it's fine for my laminate-only applications, as well as some light woodworking purposes. I hesitate to recommend a trimmer rated at less than 4 amps.

BASE DETAILS

All trimmers have a hard plastic base that rides along the work surface, but on the better-quality tools, the plastic is essentially a sub-base that attaches to a more durable metal framework, that in turn attaches to the trimmer motor. A cast-aluminum base frame is best because it allows for a more secure and longer-lasting base-to-motor attachment and adjustment. In addition, the cast-metal bases make for a more balanced tool because they provide ballast to offset the top-heavy motor. A stamped-steel base frame is acceptable, but steer clear of bases that are made entirely of plastic.

Another factor to consider is how easily the base can be removed. The easiest way to change bits in a laminate trimmer is by first removing the base, that is, if the base will come off easily. The nicest bases come off very easily by loosening a knob or moving a lever. Others require the use of a screwdriver or Allen wrench, or a knob must be removed. Easy base removal is also important if the trimmer has optional base attachments that you intend to use.

DEPTH ADJUSTMENT

Many laminate-trimming bits (bevel and no-file bits in particular) must contact the laminate at a precise height to work properly, so most trimmers have some mechanism (usually a hand-operated turnscrew) that allows the user to make micro-depth adjustments. However, some trimmer bases slide freely, and depth is visually indexed with the aid of a built-in scale and line indicator mark. The micro-depth adjustment feature is handy even if you don't use a bevel bit.

BIT CHANGING

Router bits are usually changed with two manufacturer-supplied wrenches: one wrench holds the spindle shaft while the other loosens the collet nut. However, some laminate trimmers have a spindle lock button and require only one wrench for changing bits. I think a spindle lock is a nice feature, but some people find that using two opposing wrenches is easier because it gives more leverage. In any event, having a spindle lock will usually allow the user to make bit changes with a regular wrench if the supplied wrenches are misplaced, and that's rarely possible when opposing wrenches are needed.

AVAILABILITY OF PARTS AND REPAIR

Sooner or later your new trimmer is going to need replacement parts: typically a new cord, a switch, brushes or bearings. Whether you attempt to make these repairs yourself or send the tool out for someone else to fix, you're going to want quick service. I have found (at least in my area) that there is a world of difference between brands when it comes to service. Many times I have opted to buy a tool of slightly lesser quality because I know from experience that my supplier can get parts, or send the tool off for repair, and I'll get what I need back in about a week, while with other tool brands I'll be left waiting and wondering for close to a month.

Laminate trimmers typically turn at a few thousand more rpms than larger routers, so they can make smoother cuts in plastic laminate, which is thin but dense. All laminate trimmers come with an auxiliary edge-guide attachment that, when fastened to the base of the trimmer, allows the fabricator to trim laminate off flush with the edge it overhangs. All laminate trimmers have a depth-of-cut control, and some also have a micro-adjustment feature for further tuning.

There are many brands of laminate trimmers on the market, and generally speaking, all work well. The sidebar on the facing page discusses several key features to consider when you're shopping for a laminate trimmer.

Special-purpose bases

Some manufacturers (notably Porter-Cable, Bosch and DeWalt) offer optional special-purpose bases that attach to the trimmer motor in place of the standard bases. Most of these can be purchased individually with or without the trimmer motor, or as part of a trimming kit containing one motor, several bases and a carrying case. The most popular of these attachments is the offset base; two other useful offerings are the tilt base and the vacuum base.

Offset base While a standard trimmer base will trim only to within about 3 in. of a vertical surface (like a wall) the offset base (see the photo above right) allows for a closer cut, typically ½ in. An offset base makes it possible to trim laminate in places that couldn't otherwise be reached, such as on the top of a ¾-in. thick backsplash that's in place against a wall, or on the return ears of a laminated-in-place window stool. An offset-base trimmer is the standard trimming tool for some laminate fabricators.

An offset base allows cuts as close as ½ in. away from a vertical obstruction. *Photo by Rich Ziegner.*

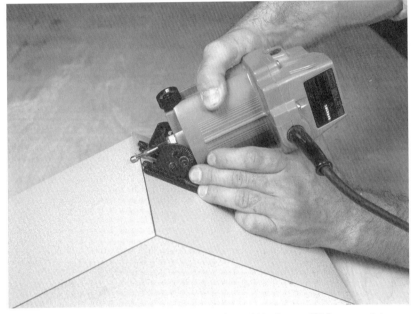

A tilt base lets you trim angles other than 90°. Some tilt bases set to obtuse angles only, while others will do obtuse and acute angles. *Photo by Rich Ziegner.*

Tilt base With a tilt base (see the photo above), it's possible to trim edges that meet at angles other than 90°. I used to own a tilt-base attachment as part of a trimming kit, and never once used it. For most countertops, a tilt base is not a necessity. I've never made a countertop where I absolutely needed a tilt base.

This vacuum-base attachment from Art Betterley Industries sucks up the the dust and chips created when plastic laminate is trimmed.

Vacuum base Another notable base attachment is the vacuum base made by Betterly Industries (see the photo at left). The vacuum base is screwed on in place of the standard plastic sub-base, and the hose is connected to a shop vacuum. The Betterly vacuum base does a commendable job of sucking up all those chips and dust that trimming laminate generates, and it will fit Bosch and Porter-Cable trimmers.

TRIMMING WITH AN EDGE GUIDE

Edge guides allow for precise edge trimming with a ¼-in. straight-cutting router bit. The typical edge guide has a roller bearing positioned just below the bit; you adjust the bearing so that, as it rolls along the edge, the bit will trim the overhanging laminate off flush. There are two fundamental problems with using edge guides. First, and most notably, it takes a while to adjust them properly; and second, they will trim off only about 1 in. of overhanging laminate at a time. I once asked the national sales representative for one trimmer maker if he thought anyone ever really used the edge guides. He frankly admitted that he didn't, at least not any of the professional fabricators he knew. I suspected that was the case because I've never used them myself and don't know of anyone else who has.

Nevertheless, for my magazine review of laminate trimmers (see the sidebar on p. 86) I did make it a point to try out the various edge guides, and I can tell you that, despite their shortcomings, when adjusted properly, they all do the job they are designed to do. I even found advantages to using an edge guide: Since the pilot bearing is detached from the trimming bit, it turns independently of the cutter, and it will therefore never gum up with glue, seize and rub a burn mark in the laminate, as bearing-piloted bits are wont to do on occasion. Straight-cutting bits generally cost less than special laminate-trimming bits.

TRIMMING WITH PILOTED BITS

The more popular alternative to edge guides are specially made laminate-trimming bits with pilot bearings on the ends. The pilot bearings on flush-trim bits are precisely sized to eliminate the need for overhang adjustment. Consequently, I can chuck a laminate bit in my trimmer and be using it in a fraction of the time it would take to install and adjust an edge guide. What's more, my trimming progress is not limited by an oversize overhang, because the bit will zip through any amount of excess laminate and get right to work at the edge.

There are many different kinds of bearing-guided laminate trim bits, but they can all be divided into two categories: self-piloted bits and roller-bearing guided bit. Available bit styles include flush-cut, bevel and no-file.

Self-piloted bits

The simplest trim bits have a self pilot (see the photo at left on the facing page). They are made out of one piece of solid carbide steel; the pilot bearing is milled into the end, below the cutting edge.

Since the bearing is really nothing more than an extension of the bit's shank, it spins at 20,000 to 30,000 rpm right along with the cutting tooth. To prevent marring or friction burns when the pilot spins against the laminate, the laminate must be lubricated with petroleum jelly or with a soft wax stick or aerosol lubricant made specifically for this purpose (see the photo at right on the facing page). (Lubricant is not necessary, and should not be used, for cuts where the bearing rides against bare underlayment.) Even with the lubricant though, it's prudent, when using a self-pilot, to keep the trimmer moving forward while

Self-piloted trim bits are milled out of a piece of solid carbide steel. The pilot is an integral part of the bit and spins with the cutter.

To keep self-piloted bits from marring a laminate edge, spread lubricating wax on the laminate guide surfaces prior to trimming. Petroleum jelly or a specially made aerosol lubricant can also be used.

the pilot is spinning against the edge; pull it away from the edge if you stop moving the tool.

Self-pilot bits work well and are popular with many professionals because they're inexpensive and durable, and the bearing is maintenance free. However, self-pilot bits have only one cutting edge, and therefore the cut they make is not as smooth as that made by other trim bits. Also, self-pilot bits are suited only for trimming laminate.

I'm a firm believer in keeping things simple, so I recommend you use (or at least try using) the flush-cut self-pilot bit. It's inexpensive, will handle all your trimming needs, needn't be adjusted precisely, and has no bearings to fuss and worry over. If laminate guide edges are lubricated, self-pilot bits perform flawlessly. The fact that they don't cut as smooth as other kinds of bits is, in most instances, unimportant because their cut isn't that bad, and most edges will still need to be finish-filed.

Bearing-guided bits

Laminate trim bits with a roller-bearing pilot are more common than self-piloted bits, and I suspect this is in large part because many users feel more comfortable having a bearing that rolls along the guide edge, rather than spinning against it. But there are other advantages too.

Bearing-guided bits have a wide tooth surface that is well suited to various woodworking operations. Flush-trim bits in particular are ideal for some template-cutting situations, though the template will have to be placed on the bottom of the workpiece.

Most bearing-guided bits have two carbide-tipped cutting teeth and therefore leave a cut that is smoother than that made by a self-piloted bit. Three-

Contact cement is nasty stuff to tangle with, and roller-bearing pilots that are not maintained will eventually gum up with a rubbery glue residue. Some fabricators wait until the bearing is completely clogged with glue, then toss the bit in a jar of solvent to soak. In the meantime, they buy another bit, and eventually end up throwing out the old one. There is a better way.

The trick to keeping a bearing rolling along dependably for a long time is to clean it often. Keep an old toothbrush, some solvent and a rag on hand to do the job. After cleaning, it's important to lubricate the bearing with router-bearing lubricant, which is available from almost any mail-order tool catalog that sells bits. If you take a minute to brush off and lubricate the bit whenever it starts to get a little dirty, the glue build-up won't get ahead of you.

tional chipping cut. For the smoothest possible cut, keep the bearing scrupulously clean (see the sidebar at left).

Roller-bearing pilots made of steel are the norm, but some fabricators prefer to use bearings with a tough plastic (Delrin) sleeve over them because the plastic is less likely to mar the laminate than a metal bearing. I used a Delrin sleeve for the first time while writing this book, and I liked it quite a lot. I think a Delrin sleeve would prove very useful when working with gloss-finished laminates, because such surfaces are particularly susceptible to marring.

Flush-trim, bevel and no-file bits

Laminate trimming bits (both self-piloted and bearing-guided) are available in flush-trim or bevel-cut styles (see the photos below). Flush-trim bits must be used when trimming laminate pieces flush with the underlayment (i.e., edge strips), and they are also perfectly suited to trimming top pieces, where the pilot will ride along a laminate edge below it. In either case, it's important to know that flush-trim bits cut almost, but not

and four-tooth bits are available for an even smoother cut. For extra smooth cuts, it's also possible to get bearing-guided bits with helical cutters. Helical teeth wrap slightly around the body of the bit, and therefore do their job with a shearing action, rather than a conven-

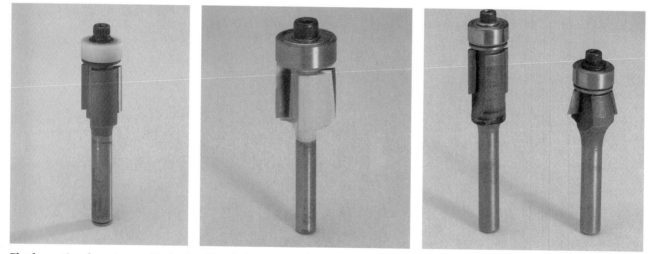

Flush-cutting bearing-guided trim bits (left and middle) come in various styles. Bearing-guided bits with a bevel cutting edge (right) take a while to adjust for precise depth of cut, but they can help eliminate hand filing.

If a flush cutting bit is set too low, and the trimmer is tipped a bit while trimming, the cutter will shave into the face of the laminate.

perfectly, flush; they are supposed to leave a slight overhang (a few thousandths of an inch), which must then be filed off—more about this shortly. Setting the depth of cut for a flush-trim bit is not critical, but it's best to adjust the bit so that no more than ⅛ in. of cutting tooth extends below the sheet being cut. If the bit is set deeper, and the trimmer base is inadvertently tipped while making a cut, the blade will slice into the face of the laminate (see the photo above).

Bevel bits are typically used only for finish trimming the top sheet of a countertop assembly. They work best if the edge is first flush-trimmed and all glue and trimming residue is removed. Bevel angles range from 7°, which is standard on all self-piloted bits, to 25°. The greater the angle, the wider will be the dark stripe of phenolic backer that shows at the edge (see the photo at right).

An edge trimmed with a bevel bit will show a wider stripe of dark phenolic backer (top sample) than an edge trimmed with a flush-trim bit (bottom sample).

Adjusting the depth of cut is critical when trimming a countertop's top sheet with a bevel bit.

Too high
A cut adjusted too high leaves an overhang on the top sheet.

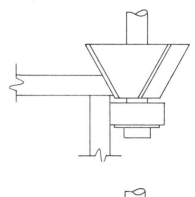

Too low
A cut adjusted too low lops off part of the edge strip.

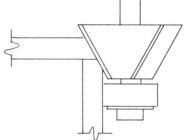

Just right
A cut adjusted just right forms a crisp edge at the glue line.

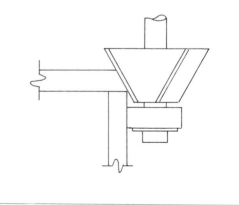

With a bevel bit, depth of cut must be set very precisely (see the drawing above) because the laminate edge to be beveled is only about 1/32 in. thick; if the bit is adjusted too high, it won't trim the top sheet back far enough, and if set too low, it will cut into the decorative face of the laminate edge strip. When using a bevel bit, start with the cutter adjusted high, and lower it in small increments, making a test cut at each setting, until the desired bevel is attained. This is where a micro-depth adjustment feature on the laminate trimmer proves itself very useful. A carefully set bevel bit can eliminate a lot of finish filing because it can be set to trim off the overhang, leaving only some milling

This trimmer with self-piloted bit is being directed around the edge strip in a counterclockwise direction, which is the proper direction for trimming operations other than around countertop cutouts.

marks to file smooth. Just don't make the mistake of using the bevel bit for flush-trimming edge strips.

Another bearing-piloted bit worth mentioning is the no-file bit. Instead of beveling the edge, the no-file bit rounds it over. These bits are adjusted like bevel bits but the depth of cut is not as critical. Like bevel bits, no-file bits are best used after flush-cutting an edge and cleaning off all glue residue. No-file bits work quite well and do eliminate the need for virtually all filing, but the rounded edge reveals considerably more dark phenolic backer than any other trim bit does.

Trimming in the proper direction

A router bit, when directed into the workpiece, spins counterclockwise. For most of your laminate trimming, you will want to direct the trimmer around the edges in a counterclockwise direction, so the bit turns into the edge it's cutting. Think of the surface that your trimmer's base is resting on as an odd-shaped clock face and move the tool accordingly (see the photo above). It's easy to remember what to do: Think **counter**clockwise on a **counter**top. However, when routing around a cutout opening, as for a sink, move the tool in a clockwise direction.

Nothing scares a laminate fabricator more than the thought of bearing loss and bit creep. Both are relatively rare, but not unheard-of, occurrences that can be prevented.

BEARING LOSS

Bearing loss is when the bearing pilot falls off during use. Bearings are typically held on with a small screw or a nut, and normal tool vibration sometimes loosens the connection. The result is an uncontrolled and devastating gouge in the edge of the workpiece. Prevention is simple: Make sure the bearing is securely attached before using the trimmer. Snug-check it with the appropriate wrench. Never assume that brand-new bits are ready to roll right out of the package; I know of one fabricator who was using a new bit to mill a detail in a countertop wood edge, and the bearing fell off in the first few feet of use. (Of course, self-pilot trim bits will never let you down in this regard.)

BIT CREEP

Bit creep happens when the bit slips out of position as you are using it. With a flush-trimming bit, some creep could happen without any adverse affect, but with a bevel-trimming bit, where precise adjustment is critical to success, the consequences would be disastrous. There are three reasons for creep, and to understand the phenomenon better, you first need to understand how router bits are held in place.

The part of the router that grips the bit is called the collet (see the drawing at left). There are various collet configurations, but the bet-

ter designs consist of a cone-shaped split-ring collet insert that fits into a tapered pocket called the arbor. The arbor is an extension of the tool's motor shaft, and it is threaded around the outside to accept a locking nut, When a bit is slipped through the locking nut into the collet and the nut is then tightened down, the split-ring collet insert clamps securely onto the router bit's shank.

If not enough, or too much, shank is inserted into the collet, the bit can come loose. Ideally, the bit should be gripped by at least the entire length of the collet, and this distance will vary because collets vary in their size. A bit should never be inserted into the collet all the way to the underside of the cutter because the transition between the cutter and the shank is tapered, and tightening onto the taper will not provide a secure hold.

Dirt or rust on the collet insert, the inside surface of the arbor or the router-bit shank can also cause the bit to come loose. Each of these parts should be shiny clean, and to keep them that way, you should periodically clean them with fine steel wool or a Scotch-Brite type abrasive pad. Don't use any lubricants or sandpaper for cleaning.

A worn-out collet insert, which is an inevitable consequence of using the tool, will not hold the shank securely, and the bit may come loose. The collet will wear more at its top and bottom, leaving only the center portion with good gripping ability. Although collet wear occurs more quickly on routers put under heavy wood-cutting loads, it's still a factor to be reckoned with on a laminate trimmer. The best prevention is to buy a new collet if the old one starts to give you trouble.

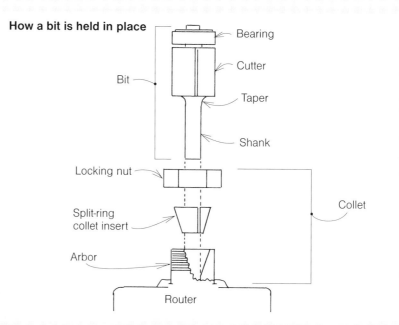

How a bit is held in place

Bearing

Cutter

Bit

Taper

Shank

Locking nut

Split-ring collet insert

Arbor

Collet

Router

This gouge in the laminate edge was caused by an irregularity in the path of the pilot bearing when the bottom of the edge strip was trimmed.

Ensuring a trouble-free job

Avoiding problems is always better than fixing mixtakes. So install your bits carefully (see the sidebar on the facing page), take the time to set up properly, make plenty of test cuts until you're satisfied with the results, and concentrate 100% on what you are doing.

If you elect to put your edge pieces on with the laminate overlapping at the top and bottom (see pp.77-78), make sure there are no surface imperfections along the pathway of the pilot bearing before trimming off the bottom edge. Gaps between build-up strips, or countersunk screw heads, will interfere with the smooth travel of your pilot bearing and ruin the edge strip (see the photo above). You can check for trouble spots visually or by sliding your fingertips along the underside. If you detect a depression, fill it with body filler (see p. 41) or put a reference mark on the edge and freehand the bit past that area, leaving a small bump of excess laminate. This can be flattened with a file, a sanding block or a sharp block plane, as described in the pages that follow.

If you're going to work with laminate, a sharp laminate file is a necessity for smoothing down milling marks and carefully removing the slight overhang left after flush-trimming. A laminate file will cut laminate much better than files made for wood or metal. You can get coarse- or fine-cutting laminate files, with the same cutting surface on both sides, or you can get a file that's coarse on one side and fine on the other. Most laminate files also have teeth cut into the edge of the blade (see the photo below right).

When new, a laminate file is very sharp and remarkably efficient at removing laminate, but it will dull with use and after making a few countertops, you'll need another file. Laminate files are sold by plastic laminate suppliers, and by the mail-order tool suppliers listed in Resources on pp. 127-129.

Laminate files made of solid carbide steel are also available, and these are purported to last 50 times longer than conventional files. Although they cost about six times more, they are obviously worth it if the tool does last that long. (I can't personally vouch for the longevity of a carbide file, but I don't doubt the claim.) When dull, carbide files can be reground by the manufacturer at a fraction of the cost of a new file.

I once tried out a carbide file. It was well made and it worked nicely, but the tool was difficult for me to get used to. The file consists of a ¾-in. by 6-in. piece of carbide mounted to a plastic handle, and there is less cutting area available to work with than on a regular laminate file. Also, the carbide cutter has no teeth on the side, and not having that cutting surface made me realize how useful it really is when finish-filing edges.

When cutting with a laminate file, always move the tool so that you are cutting into the decorative face of the laminate first, and file toward the back of the piece. If you file in the opposite direction, you might chip out the face, or stress and break the adhesive bond. Cut only on the forward stroke, then lift the file (or let off downward pressure), reposition and cut on the forward stroke again.

A clean file cuts much better than a dirty one, so you will want to clean the file often. For best results, use a file card, which is a stubby wire brush made specifically for cleaning files.

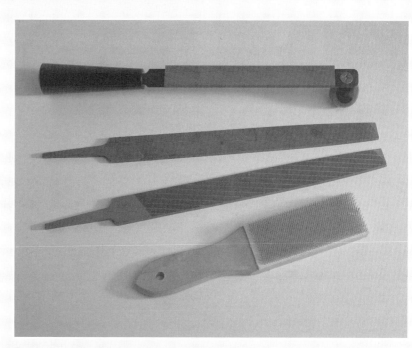

Equipment for filing plastic laminate. From the top: a carbide file, a coarse-cutting file, a fine-cutting file and a file card.

After trimming the laminate flush to the underlayment, smooth away the rough milling marks by filing with the file held perfectly flat on the countertop.

SMOOTHING THE TRIMMED EDGES

Trimmed laminate edges are not as smooth as they seem at first glance; a closer inspection will reveal that the cut is actually quite bumpy. The irregularities result from milling marks normally left by the cutter and because the pilot bearing typically rides over dust and chips as it rolls along. A laminate edge that will be overlapped by another piece of laminate (i.e., the top of edge pieces, or one side of an outside corner) therefore need some further work.

To smooth the edge, lay a coarse-cutting laminate file (see the sidebar on the facing page) perfectly flat on the underlayment and make a few leveling passes (see the photo above). Holding the front half of the file down firmly to the underlayment while cutting with the back half of the file blade will ensure that you don't round over the front edge.

The objective when filing an edge flat to the underlayment is not to file it down perfectly flush, but to level all the little milling bumps. A few long, easy swipes with a sharp file should do the trick. If

A pad sander held flat to the surface can also be used to smooth out the milling marks in a trimmed edge. Don't oversand if you use this method.

you watch the edge closely, you'll see the milling irregularities level out, and you'll know when to stop.

Trimmed edges can also be flattened using a pad sander (see the photo above). The sander is particularly useful for smoothing bottom edges, where a file is awkward to use.

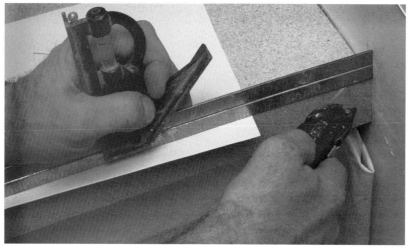

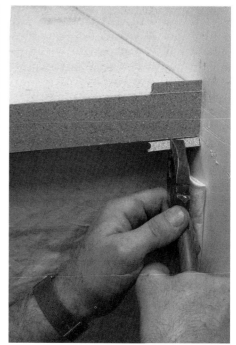

Untrimmed sections of laminate the trimmer couldn't get to are called 'ears.' They can be removed by first scoring with a utility knife and straightedge. Elevate the straightedge slightly by placing it on a piece of thin cardboard, and score lightly to start (above). After scoring, you can snap off the ears with a pair of pliers (right), and then smooth down the edge with a file.

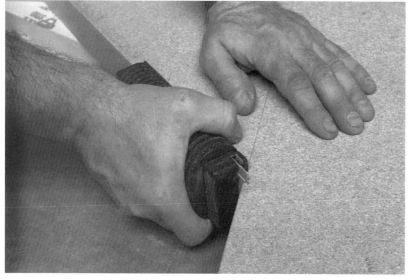

A carbide-tipped scoring tool precisely and effortlessly scores laminate ears.

When site-building a countertop that abuts a wall and can't be pulled away from the wall for trimming, the trimmer bit won't be able to cut all the way to the wall, because the base will hit first. The laminate "ears" that remain can be removed by scoring with a utility knife guided along a straightedge and snapping them off with pliers, as shown in the photos above. Another way to remove laminate ears is with the special scoring tool (available from laminate suppliers or by mail order), shown in the photo at left. Top-sheet ears can also be removed with either of these techniques.

FINISH-FILING THE EDGES

After the top sheet is glued down and trimmed flush, the edges can be finish-filed. Your primary objective here is to remove the slight overhang left by the trimmer bit (assuming you used a bit that leaves an overhang) so that the top edge is perfectly flush with the decorative face of the edge laminate, but not to file into the decorative face of the edge strip. Finish filing is a skill that is acquired; the more you do it, the better you get. But the process isn't really that difficult, and there is no reason why a person with little experience can't do a good job too.

While some fabricators clean the edges with a solvent-dampened rag before finish-filing, I prefer to leave the typical residue of glue, lubricant and phenolic pieces on the edge. As you'll see shortly, this edge junk gives me a visible filing reference, though it means I'll have to clean the file more often to keep it cutting well.

Use a smooth-cutting file when finish filing. When filing edge strips that meet at an outside corner, hold the file almost flat to the edge (see the photo at right). You can hold a precise angle by gripping the end of the file with your fingertips, so they ride against the edge and prevent the file from contacting the decorative laminate surface. When finish filing top edges, angle the file as shown in the photo below right and stroke in a forward, downward motion.

As you get nearer to the final flush-swipe, you may want to change technique to gain more control: Cut by moving the angled file ahead but not down, and hold your finger between the file and the edge at the bottom to help maintain a fixed angle. You'll know when you've reached perfect flushness because any glue and other residue will shave cleanly away from the glue line. When this happens, don't do another stroke in that area. Concern yourself only with filing away the residue at the seam; the rest can be cleaned off with solvent. When in doubt about whether the overhang has been filed sufficiently flush, slide your fingernails up the edge the edge strip. If they catch on the top sheet, there is still overhang to be removed.

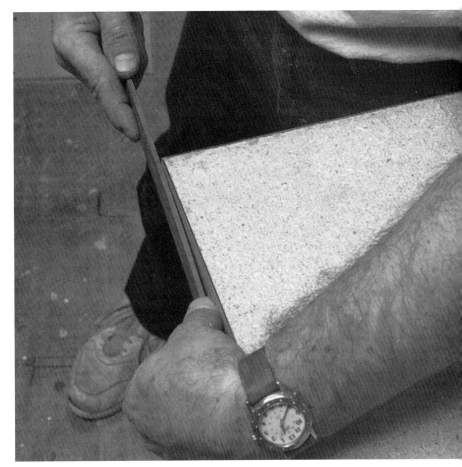

Maintain a precise, almost flat angle for finish filing outside corners by gripping the end of the file with your fingertips, holding the file face off the laminate surface slightly; stroke forward with your fingers indexed against the laminate.

When finish-filing a top edge, hold the file almost flat to the edge (note the space in the photo), angle the bottom ahead slightly, and file in a forward and downward motion.

Soften the sharply filed edges with a few light passes of the file (you can also use a piece of fine sandpaper).

At inside corners, the trimming bit will leave a small radiused overhang of laminate. To file the corner square is tempting but unwise because it invites stress cracking; leave it the way it is.

After you've finish-filed a flush-cut edge, the top corner will be very sharp and it should be softened with a few light passes of the file (see the photo at left). A light sanding with 100-grit sandpaper will also do the job.

If the bottom edges of edge strips are a bit rough, they can be smoothed with 100-grit paper in a sanding block or with an electric pad sander. Then, just as on the top edge, knock down the sharp outside part. A file can also be used to smooth the bottom, but sanding is easier if you're working with the countertop in a top-side-up position.

With an edge that has been bevel-trimmed, the finish-filing objective is to make a few light passes over the bevel to smooth out any milling marks. Also, bevel bits will sometimes not cut a consistent bevel and you may need to flush-file any areas of overhanging laminate. When in doubt, use the fingernail test.

Avoiding stress cracks at inside corners

When filing the top sheet of a countertop with inside corners, don't remove the small radius left by the trimming bit (see the photo at left). It's tempting, while finish-filing along the edge, just to keep on going into the corner and file it square, but if you do this you are inviting a stress crack to form.

I have to confess that in the past I didn't take stress cracking seriously and handled inside corners somewhat arbitrarily; sometimes I'd leave the radius, and sometimes I'd file the corner square. It wasn't until I got a call from a customer telling me she had a crack in the corner of her countertop that the importance of leaving that little curve of laminate really hit home. I had made the top a year earlier. Why that particular corner chose to crack, and other counters I had filed square didn't, I'll never know.

Fortunately, that particular crack was very fine and only about 3 in. long. But I knew that once started, the crack would lengthen, much like a small crack in a car windshield. The only thing I could do, short of replacing the entire counter, was to drill a ⅛-in. hole through the laminate at the very end of the crack. I hoped this drastic move would relieve the stress and stop the crack from growing. To patch the hole I mixed tubes of colored filler made specifically for countertops (see the photo at right). To my relief, the customer agreed to the experiment, the patched hole didn't look so bad, and, best of all, the crack never got any worse. It was a close call, and another lesson I learned the hard way.

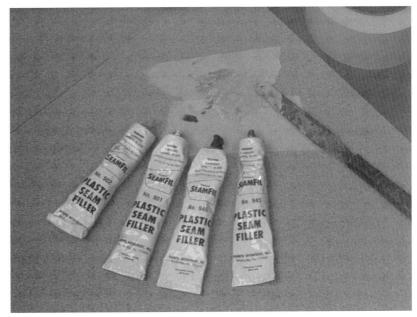

Tubes of colored seam filler can be mixed to get a respectable (though not perfect) color match with most laminates. You can use the filler to repair poorly matched seams and surface nicks, but it will not work on scratches.

INSTALLING THE COUNTERTOP

A finished countertop must be fastened securely to the cabinets below. This is best done by driving drywall screws up through the cabinet corner blocking into the build-up strips. Drill a clearance hole through the corner blocks and the job will go much easier. Of course, you'll want to make sure your screws are long enough to find good anchor in the underlayment, but not so long that they go up through the countertop. I've seen guys drive screws through the counter on a couple of occasions, but in those instances the counters were post-formed and lacked adequate build-up to screw into, so the installers were attempting to anchor a screw into the ¾-in. deck sheet. If you used 3-in. wide build-up strips on the underlayment, there shouldn't be any problem finding build-up to screw into. And if you size your screw length to go ¾ in. into the build-up, you'll make a solid connection without the screw tip ending up anywhere near the top surface.

With the countertop fastened in place, take a close look at your work. If you notice a seam that didn't go together perfectly, or an unfortunate chip some-

where in the laminate, you can use colored filler to match the laminate. For most laminate styles, you'll need to get two or three tubes of color and mix them together as directed in a chart. The filler dries fast and hard, and will do a decent job at hiding many flaws, although it does not fix scratches.

When all is done, give the counter a final cleaning. Use solvent to loosen adhesive spots. Stubborn marks of unknown origin can usually be removed with a well-diluted bit of Soft Scrub type abrasive cleaner. Countertop polishes are also available (see Resources on pp. 127-129). Although polish is not normally needed on a brand-new countertop, I sometimes leave a bottle for my customers as a parting gift.

6

BACKSPLASHES

A 4-in. high attached laminate backsplash with ceramic tile above is one of many backsplash design options.

Many people consider the backsplash an opportune place to show their flair for design, and it is. But don't forget that a backsplash has practical functions as well. First, the backsplash must be a durable barrier that prevents objects on the countertop from being shoved against the back wall and damaging its surface. Second, the backsplash acts as a seal against liquids seeping behind the counter. Third, the backsplash allows for seasonal dimensional movement between the counter and the wall framing without showing gaps or destroying the integrity of the liquid seal. And fourth, a backsplash should be easy to clean, especially in areas most susceptible to spatters, such as behind a stove or sink.

With those criteria in mind, I believe the best backsplash for a kitchen countertop is either ceramic tile or plastic laminate applied to the entire wall area between the countertop and the bottom of the wall cabinets. A more economical, and still perfectly functional, backsplash option is a 4-in. high attached backsplash of plastic laminate in conjunction with a piece of laminate, or a section of tile, on the wall area above and behind the stove or sink (see the photo at left). It's also possible to have a wood backsplash, but I don't recommend it (see the discussion on p. 109).

Whatever option you select, it must be installed properly if it is to serve its functions. In this chapter I'll show you the correct way to install a full-wall laminate backsplash and a 4-in. high attached laminate backsplash. For detailed information on installing a ceramic-tile backsplash, I recommend Michael Byrne's book *Setting Tile* (The Taunton Press, 1995).

ATTACHED BACKSPLASHES

Attached backsplashes are the most common type of laminate backsplash. They are usually about 4 in. high (though this measurement can vary) and they're made by gluing laminate to a ¾-in. thick underlayment material, usually particleboard. Constructing backsplashes is a basic exercise in plastic-laminate fabrication; they are made similarly to the countertop, with the end pieces going on first, followed by the front face, and finally the top. If need be, assembled backsplash pieces can be trimmed to length with a chopsaw or ripped down with a table saw.

To make long lengths of backsplash you may need to join pieces of underlayment. This can be done with dowels and glue, or if you have a pocket hole cutter, use that with the appropriate screws (fill the pockets with auto-body filler before laminating). Such connections are not overly strong, but will be strong enough if you are careful not to put undue stress on the joint when you move the piece around.

If two pieces of laminate must be seamed together to cover a long length of backsplash, keep the seam as far away from the joint in the underlayment as possible. It's also a good idea to try to locate backsplash seams in an obscure place, such as near an inside corner; directly in front of the sink would not be good.

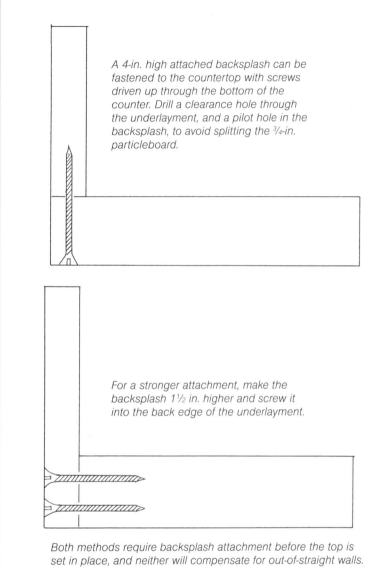

ATTACHING A BACKSPLASH WITH SCREWS

A 4-in. high attached backsplash can be fastened to the countertop with screws driven up through the bottom of the counter. Drill a clearance hole through the underlayment, and a pilot hole in the backsplash, to avoid splitting the ¾-in. particleboard.

For a stronger attachment, make the backsplash 1½ in. higher and screw it into the back edge of the underlayment.

Both methods require backsplash attachment before the top is set in place, and neither will compensate for out-of-straight walls.

Attachment methods

There are several ways to install an attached backsplash; opinions vary as to which is best.

Screws The backsplash can be screwed on before the countertop is set into position, as shown in the drawing above. Screws can be used only with shop-built countertops that have straight back edges. This method can not be used to make the backsplash conform to wall irregularities.

For short countertop sections such as vanity tops or with simple counter configurations that abut relatively straight walls, a screw-attached backsplash may be ideal. Minor wall irregularities that show as gaps at the top of the backsplash can often be eliminated by feathering a layer of joint compound down the length of the backsplash. If the back wall is already finished off, some fabricators will install the countertop with the top edge piece off the backsplash, then scribe the piece to the wall, glue it on and finish-trim it in place. (For information on scribing, see pp. 60-62.)

Silicone caulk Many fabricators opt to apply a backsplash after the counter is installed by applying a bead of silicone caulk to the back and bottom edge of the backsplash. The backsplash is then pushed into position and held in place with an assortment of clamps, sticks and wedges until the caulk cures. No clamping setup is needed if hot-melt glue is dabbed about every 4 in. or so on the back of the backsplash before setting it in position; the hot-melt will lock the piece in place until the stronger and more flexible silicone has a chance to set up.

The fabricators I know who use this technique think silicone is the best way to go, but I've never felt comfortable with it. On some installations I've seen the backsplash and countertop gape apart, and I've also seen damage from water that managed to seep under the backsplash, wick up into the particleboard, and wreak havoc with the underlayment and adhesive bond.

Backsplash clips To my mind, the hands-down best way to attach a backsplash to the countertop is with clips specially designed for the purpose; I've used these little gems for several years now and wouldn't even consider using anything else. The clips, called Smart Clips, are distributed by D & K Sales (see Resources on pp. 127-129). With the Smart Clip system, you can attach the backsplash after the countertop is fit in place, and the clips will allow the backsplash to conform to minor wall irregularities. The connection is secure and permanent, yet you can detach the backsplash if the need arises. And it's almost impossible for liquids to get past the backsplash seal and wick up into the particleboard underlayment. The only disadvantages to using the clips are that they cost more than a few screws or a tube of silicone caulk and they take longer to install. But the extra cost isn't unreasonable, and I believe the time and money are well spent.

At the heart of the Smart Clip system are the little plastic clips that get screwed down to the countertop every foot or so along the wall where the backsplash will go. The clips are about ¼ in. thick, and they're hidden from view in a rabbet cut in the back side of the backsplash (see the photo on the facing page); the cut removes all the underlayment on the bottom length of the backsplash, leaving only a ⁹/₃₂-in. lip of laminate hanging down. Screws are driven up into the bottom of the backsplash opposite each clip, and their heads are left protruding ⁷/₃₂ in., just enough to clip securely into a slot in

The Smart Clip system: Screw heads protruding from the rabbeted underside of this backsplash mock-up (seen here from the wall side) lock into plastic clips at the back edge of the countertop. A bead of caulk seals the joint.

each clip as the backsplash is pushed into place. The edge of laminate that hangs down blocks the clips from view, and is surprisingly sturdy.

Before the backsplash is snapped into final position, a bead of caulk is applied along the countertop in front of the clips. The caulk acts as a dam. Any liquids that might make it past the tightly clipped backsplash will be stopped at the caulk barrier, and since the underlayment is completely routed away from the bottom edge of the backsplash, there is no material there for the moisture to wick into and ruin.

A Smart Clip backsplash requires moderately heavy pressure to snap it into place. To make the job easier, I mark the placement of the clips with a pencil on the countertop, and then, starting

on one end, apply pressure at each clip location by hammering on a ¾-in. thick board held tight to the bottom of the backsplash. Since the board is thicker than the lip of laminate, no harm is done.

As you might guess, a system like this requires precision for all the components to fit together properly. The rabbet has to be the right size, the screws in the backsplash have to be positioned correctly relative to the thickness of the backsplash, and they must protrude just the right amount. You could do this work with basic woodworking tools and some careful attention to the details (the instructions are clearly illustrated), but if you were going to do more than a couple of countertops, the Smart Clip installation kit would be a good investment. The kit consists of a specially designed ½-in. shank rabbeting router

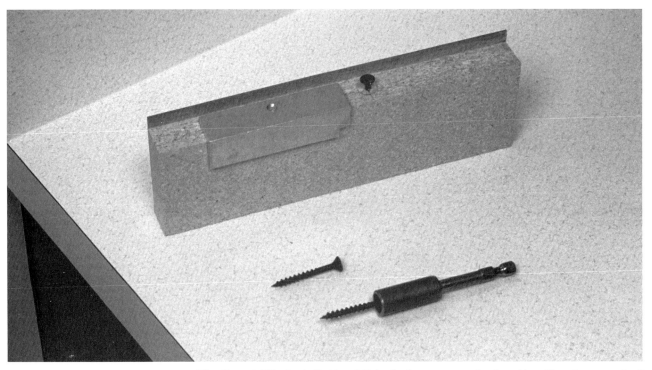

The Smart Clip installation kit includes a gauge for locating the placement of the screws in the bottom of the backsplash and a screwdriver bit for setting each screw at the required depth. A rabbeting router bit (not shown here) also comes with the kit.

bit, an angled aluminum gauge to help with proper positioning of the screws in the width of the backsplash, and a screwdriver bit that leaves just the right amount of screw protruding. The installation kit (see the photo above) makes the whole process of using Smart Clips substantially easier.

Although Smart Clip backsplashes fit remarkably tight to the counter, light-colored laminates will still sometimes show a dark hairline gap in a few places along the bottom edge. The space is not a functional problem and could be left alone without any adverse effect. But my critical eye doesn't like to see such things, so I get rid of it by applying a

very fine fillet of white silicone caulk along the juncture. A small line of silicone deposited directly down the joint can be pressed into the corner and wiped smooth with a swipe of the finger without making a smeary mess.

FULL-WALL LAMINATE BACKSPLASH

Putting a sheet of plastic laminate directly on the wall at the back of the counter used to be very popular and is still done quite a bit because it has definite advantages. A laminate-covered back wall is a durable, easily cleaned and colorful surface, just like its matching countertop. But an added benefit to a full-wall backsplash is that the laminate is thin and therefore does not take up as much countertop space as a applied ¾-in. thick backsplash. Although ¾ in. may not sound like a lot, it's still something, and behind the sink in particular, that small gain will mean a more open area, which looks better and is easier to keep clean.

Applying a sheet of plastic laminate directly to a wall requires an aluminum cove strip at the wall-to-counter juncture as well as wherever two sheets of laminate meet at an inside wall corner. Outside corner strips are also available, as are cap pieces, which fit over and finish off the edges of the laminate. These trim pieces are usually available from the same places that sell plastic laminate.

Some people don't like the look of aluminum strips, and they're certainly entitled to their opinion, but I'll turn a job down before I install a full-wall backsplash without them. The strips serve a very important purpose by allowing for seasonal dimensional changes between the wall and the countertop without unsightly, moisture-vulnerable gaps opening up.

There have been a few occasions in the past when, at the customer's request, I installed full-wall backsplashes without the strips. I would put the laminate on the wall before making the counter, and make sure the bottom edge of the sheet was positioned below the countertop level. Then I'd make the countertop, cope the laminate to the wall for a good fit, and caulk the seam. In every instance, the joint eventually separated, the caulk seal failed, and the customer was unhappy. Although I'd go back and recaulk the seam, the gap would always open up again. Usually the gap was minor, but it wasn't pretty; and there is no caulk, applied as a small fillet in the

corner, that can withstand the structural stress that forms in that juncture. In parts of the country where seasonal movement is not as much of an issue, the caulk might hold up, but I still think aluminum strips are a more practical solution to the problem.

The only way I know of to build a full-wall backsplash without the need for metal strips is by putting the laminate on the wall and then Smart-clipping an attached backsplash on. Although dimensional movement will still happen, any gapping will occur at the top of the backsplash, where it is less noticeable, and the integrity of the counter seal will be maintained. This hybrid approach takes more time and effort, so it costs more. Most folks wisely opt for either the full-wall backsplash with aluminum strips or the attached backsplash.

Preparing the wall for laminate

Laminate manufacturers specifically advise that their product not be glued to plaster or drywall because there is a substantial difference between the dimensional stability of those materials and laminate, and the glue bond will fail. Be that as it may, full-wall backsplashes have been glued directly to drywall; I've torn out many of them, and I've also put several in, but I rarely do it that way any more.

On the backsplashes I've torn out, where the laminate was glued directly to drywall, the laminate was seldom if ever firmly stuck to the wall. Typically it was held in place by the bottom metal strip, switch and receptacle cover plates, window casing, and a few odd spots of adhesive. Sometimes the laminate was glued to the wall before the upper cabinets were installed, and they also helped to hold the top edge.

Despite my dire warning, sheets of laminate will continue to be glued directly to drywall in backsplash situations. If you decide to do likewise, the sheet will probably not fall off, and will therefore be adequate, but it is much better to glue the laminate to plywood.

One of the ways this can be done is to substitute an equivalent thickness of plywood for the drywall in the area where the laminate is to go. However, it is easier to glue and nail (I use construction adhesive and a pneumatic stapler) a layer of ¼-in. plywood over the existing drywall. I especially like to use the plywood overlay approach when putting a full-wall backsplash in an older kitchen. That way if I need to run wires in the wall for new receptacles, lights and switches, I can cut wherever I need to and cover up the mess with a clean, smooth layer of plywood when I'm finished. When the backsplash is glued to the plywood, the extra thickness is not evident on the countertop, or under the upper cabinets, or where it dead-ends into a door or window casing (the casings can also be notched over the plywood and laminate panel). In the few areas where the thickness does show, it can be finished off with an inconspicuous piece of wood trim.

The plywood will often need to fit around windows and under the wall cabinets. In these instances, it isn't critical that the plywood fit in place perfectly tight. Small gaps are no problem because the laminate and trim strips will cover them, and a tight fit just makes getting the sheet in place that much more difficult.

After the plywood is stapled on the wall, the next step is to install aluminum cove molding tight to the countertop. I like to caulk the wall/counter juncture with silicone and set the molding into that. Attach the strip with small nails into the wall (a plywood surface gives more to nail into than drywall), and miter the molding pieces where they meet at outside and inside corners. The metal strips can be cut with a hacksaw or with a carbide-tooth blade in a chopsaw. If you use a chopsaw, back the strips up with a scrap of wood to keep them from flexing. Feed the blade slowly into the metal, and be sure to wear safety glasses.

Fitting and gluing the laminate
Next, cut the laminate sheet to fit by taking careful measurements and using a pair of laminate snips. Another way to get the proper sheet size is by making a pattern (see pp. 62-64). I've made backsplash patterns up to 14 ft. long that jogged around windows, conformed to unlevel wall cabinets, and surrounded numerous electrical boxes. For a simple piece of laminate with only a couple of electrical-box cutouts, you can mark out the openings using a framing square to make reference marks on the countertop, then transfer the marks to the laminate after setting it in place. Remember when making inside corner cutouts (as for electrical boxes), to radius the corners with a drill bit to avoid stress cracking (see p. 51 and pp. 100-101).

When sizing the laminate sheet, the slot in the cove-molding bottom strip has a little room for leeway, and if you use a cap at the top of the sheet (the cap gives a better finished look), it too has some room for play. So when you're fitting laminate in tight spaces, you can usually cut the piece ⅛ in. or so smaller than tight, and this will help make the fitting easier. To be on the safe side, always dry-fit the laminate sheet before gluing it.

To facilitate the positioning of glued pieces, use blind slats taped vertically to the wall. Plastic mini-blind slats are more flexible than metal slats and will bend enough to be used if you have to set a backsplash piece between the counter and wall cabinets that are already in place (see the drawing at right). Seat the bottom edge of the laminate all the way along the length of the cove strip, push it back against the wall enough to be sure it's in proper position, then start removing the spacer slats and make contact, much as described on p. 80. Don't neglect to roll the laminate after installation.

If you elect to install a cap piece, it can be set or slid on over an edge before making contact, and the pressure of the contact-cement bond will hold it in its place. If you want, you can also apply a small bead of construction adhesive to the back side of the cap pieces for extra security.

WHAT ABOUT WOOD FOR A BACKSPLASH?

A wood backsplash should be used with discretion. I do not recommend wood as a backsplash material near a sink, where it will be in frequent contact with water. I've seen wood backsplashes behind sinks, and I've installed some myself. Every time, the wood looked great when installed, but a few years later it looked downright dowdy. And with continued exposure to water, wood degenerates from dowdy to gross as it blackens with mold, no matter what finish is put on it.

If you insist on wood, instead of a wood backsplash consider using wood wainscoting on the back wall with an attached laminate backsplash. That way you can get the aesthetically pleasing look of wood into the area while maintaining the functional integrity of the counter-to-wall seal. It's only a 4-in. high barrier, but it makes a world of difference.

GLUING ON A FULL-WALL BACKSPLASH

Once the sheet is properly positioned, make contact at the bottom, slide slats out the top, and seat the whole sheet.

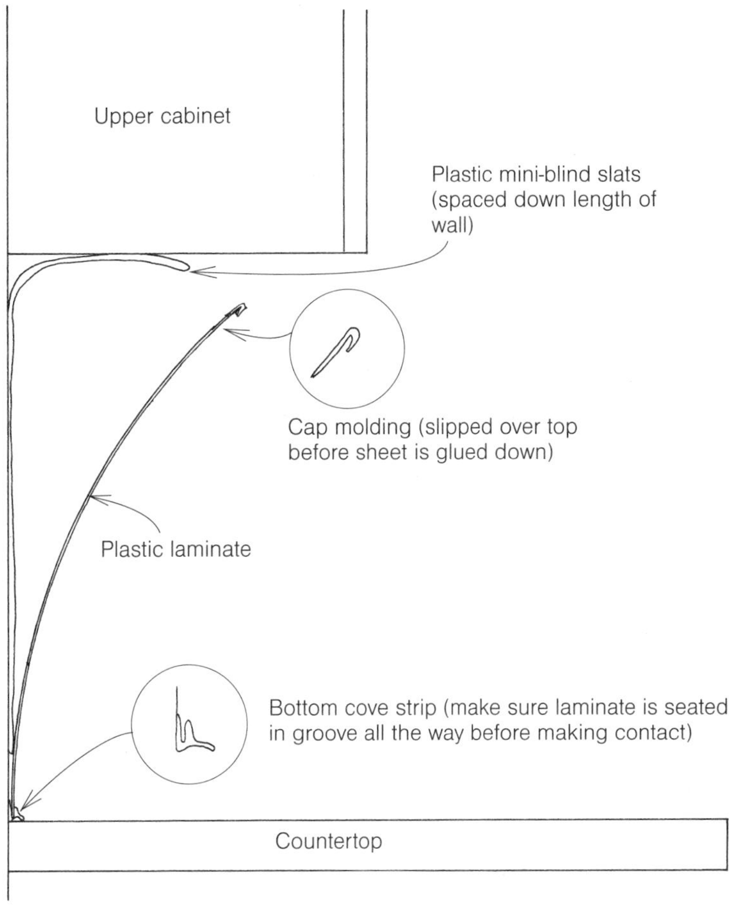

Upper cabinet

Plastic mini-blind slats (spaced down length of wall)

Cap molding (slipped over top before sheet is glued down)

Plastic laminate

Bottom cove strip (make sure laminate is seated in groove all the way before making contact)

Countertop

7

CUSTOM EDGE TREATMENTS

Thus far we've talked about the skills needed to fabricate a square-edge laminate countertop. In this chapter I want to explore the custom edge treatments that can turn a nice but ho-hum countertop into a thing of beauty. The ideas and techniques discussed here will give you the basics of custom edge construction. If you're so inclined, you can take what you learn here and create your own signature edge designs.

WOOD EDGES

Wood offers a lot of design versatility as an edge material, and it also has the advantage of being more durable than plastic laminate. Laminate is undeniably denser, but if hit with a hard object, it could chip or crack, and the spot would be obvious. Wood, on the other hand, would only dent, and the blemish will be less noticeable (it could even be steamed out). Putting a wood edge around a counter also means you won't have any top-edge finish filing to do. On the down side, wood may need periodic refinishing to keep it looking good, and some edge profiles have crevices that are not easy to keep clean (see the drawing at right). Water damage at a wood front edge, however, is not as much of an issue as it is at the backsplash.

There are two fundamental approaches to putting wood on a countertop edge. You can apply the edge strip flush with the top surface of the laminate after the laminate is glued down, or you can install the wood edge before the top sheet goes on (see the drawing on the facing page). With flush-butted edging, you end up with a seam at the top of the counter where the laminate and wood join, and I don't consider this particularly desirable from the point of view of moisture intrusion. Another drawback

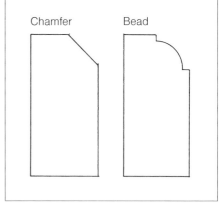

WOOD EDGE PROFILES

A simple style with clean lines, like the chamfer, will be easier to keep clean than an edge with a more involved detail, like the bead.

Chamfer   Bead

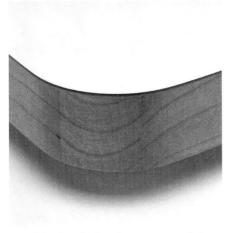

On this simple laminate-over edging, the top edge of the wood strip is left unmolded.

is that perfect flush alignment is often difficult to achieve, and planing or sanding down a wood edge that is too high is a delicate operation because the laminate surface is very susceptible to damage. The main advantage to a flush-butted wood edge is that you see more of the wood. And if flush top alignment can be achieved without the need for additional fine-tuning, the edges could be prefinished.

Laminate-over edging is the option I prefer because there is no vulnerable top seam and there is no problem with getting flush alignment; the edge is planed or belt-sanded flush to the underlayment before the top sheet of laminate is applied. After the laminate is glued down, a chamfer or other molded detail can be routed along the edge. Alternatively, the edge can be left unmolded, as shown in the photo above.

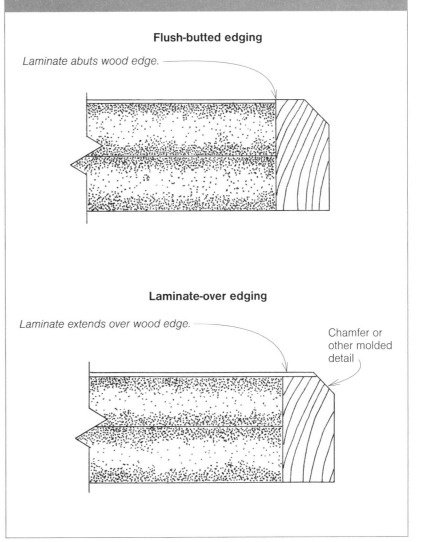

TWO DESIGNS FOR WOOD EDGES

Flush-butted edging

Laminate abuts wood edge.

Laminate-over edging

Laminate extends over wood edge.

Chamfer or other molded detail

Attaching a wood edge

There are many ways to attach a wood edge (see the drawing below) and my methods have evolved over the course of my career as I've tried new things and discovered better ways. Glue and nails driven through the face (the set nail heads are then puttied) is the most common method. This is certainly a quick and easy way to get the job done, but a row of puttied nail heads is usually obvious and detracts from the beauty of the finished product.

The edge can also be fastened with counterbored screws that are then covered with wood plugs. This approach works nicely and can look good, though the style is not suited to all decors.

A third option, which also gives you a solid connection, is to mill a groove down the face of the wood edge, and fasten with screws or nails at the bottom of the groove, then cover the screw heads with a stylish inlay strip of laminate or other material (see the top photo on the facing page). The groove can be precut in the edge with a table saw and dado blade, or it can be formed with an inlay router bit (see the photo at bottom left on the facing page). As the edges are attached, fasteners are countersunk below the line of cut before using an inlay bit.

Another "invisible fastener" edge attachment method I've used is screws driven through pocket holes in the deck and into the backside of the edge. For this approach, a pocket-hole jig is a necessity. All the pockets will have to be filled with epoxy body filler (see p. 41).

It's reassuring to attach the edging with mechanical fasteners (i.e., nails and screws), but an even coating of yellow wood glue on both edges works just fine all by itself (see the photo at bottom right on the facing page) if you hold the edging in place with 10-in. lengths of filament tape until the glue dries. Stick the tape on the bottom build-up first, wrap it up and over the front edge while

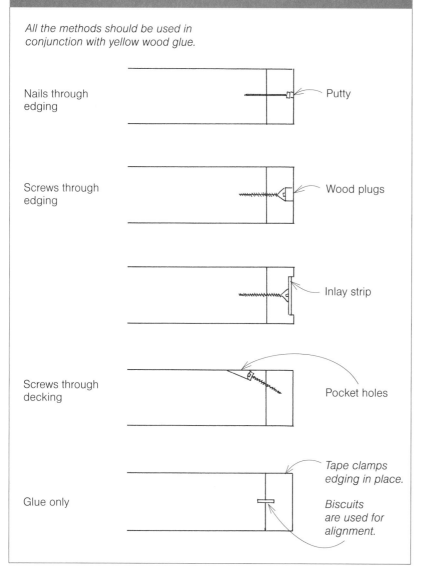

WOOD EDGE ATTACHMENT

All the methods should be used in conjunction with yellow wood glue.

Nails through edging — Putty

Screws through edging — Wood plugs

— Inlay strip

Screws through decking — Pocket holes

Glue only — Tape clamps edging in place.

Biscuits are used for alignment.

This cherry kitchen sports a distinctive cherry edge with a laminate inlay strip.

An inlay router bit can mill a precise inlay groove after the wood edge has been installed.

Glue-only edge attachments are stronger than you might think. This sample piece of underlayment with cherry edging was simply glued on, tape-clamped in place and set aside to dry. Even bashing with a 20-oz. hammer failed to break the glue bond.

To tape-clamp a wood edge, hold the stuck-on filament tape on the bottom, push in on the wood edge, stretch the tape tight and stick it down on the top. Glue should ooze out the top joint. Biscuits may be used to help with alignment.

pressing in on the edge, and stick down the rest on the deck top. Glue should ooze out of the joint (see the photo above). Even on site-built tops, where the counter can't be pulled away from the wall, I've been able to get enough tape on the bottom overhang to make a tape clamp work. To key the edge into position and keep it from falling down while tape-clamping, I use a biscuit cutter and space biscuits along the length, but I don't go to the trouble of gluing the biscuits in; the extra strength just isn't needed.

Making radiused corners

As we discussed on pp. 47-48, radiused corners on a countertop not only fulfill a practical function, but can also lend a touch of real class. This is especially true if curved edges are made of wood.

Premade outside corners can be bought in flush-butted 2-in., 4-in., and 6-in. radius pieces. To make your own, you

can laminate thin strips of wood around a form, or you could try cutting a radius out on a bandsaw. Make the curve in the countertop with a template guide and pattern-cutting bit (see pp. 48-50).

Making curves with laminate-over wood edges is considerably easier than making flush-butted edges because all you have to do is make the outside curve (see the photo at top right on the facing page) and this can be done with a template guide after the edges are on. (With flush-butted edges, you have to curve the inside edge as well.)

With a plastic laminate self-edge, tight radiuses (i.e. ¾-in.) look especially good, but a wood radius usually looks best if it is bigger. I think the 2-in. radiused wood edge shown in the bottom photo on the facing page looks very nice. Segmented curves are relatively easy to make, and there is no limit to how big you can go with them, though sizes employing one

Curved wood corners (middle) are an elegant design feature. Other options include 'chopped' corners (top) and mitered corners (bottom).

When making outside curves with laminate-over wood edges, glue on the wood in one or more angled segments (bottom), then mill the outside curve with a template guide (middle). After the laminate is glued on, rout an edge detail (top).

This kitchen has 2-in. radiused corners on the peninsula and island. A chamfer on the top and bottom of the edge lends a softer, more rounded look.

Segmented curves can be made to any size, using one or more segments. These samples show 2-in., 6-in. and 12-in. segmented radiuses.

or two segments usually look better than ones requiring many pieces (see the photo at left). To figure segment layout, draw the radius, mark the midpoint and ends of the curve, and plot out equal-size segments, keeping in mind that it's best not to remove more than half the thickness of the wood edge when template-cutting the curve.

Another variation on the laminate-over wood edge is what I've dubbed the hybrid edge, as is shown in the photo below. Here I've template-cut a tight ¾-in. radius in a wood edge (no segments needed, just a 90° mitered corner) and applied a post-formed strip

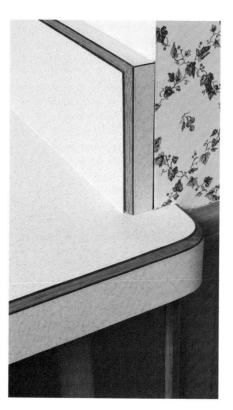

A hybrid wood/laminate edge adds a stylish touch. Here, plastic laminate applied over a wood edge was chamfered with a router bit. The same method used on the backsplash unifies the customized components.

of laminate using my iron-bending jig (see pp. 78-79). After assembly, a router bit was used to chamfer the edge.

Finishing a wood edge

Everyone has a favorite wood finish, and mine is semigloss tung oil from ZAR. It's very simple to apply, and three coats will leave a nice-looking and durable oil finish. Unlike a coating of polyurethane, the finish will not look thick, and it can be easily renewed with another coat at any time.

LAMINATE EDGES

Another custom edge option is to attach a beveled laminate edge instead of the typical squared-off shape. The bevel gives the counter more of a monolithic, or solid-surface, appearance (see the top photo on the facing page) because it is tightly mitered to the top and face pieces of laminate, so the dark phenolic stripe is eliminated. The beveled edges are actually custom-made moldings that are glued to the countertop edge after the top sheet of laminate is down. Since the bevel moldings are made to order, a fabricator can request any color or style of laminate on the bevel and face, and this opens up a mind-boggling array of color combinations and design possibilities.

Some custom laminate edges on the market have flat backs and are simply glued and tape-clamped to an underlayment edge, but the better-quality edge moldings have a ¼-in. by ¼-in. tongue on the back that is designed to fit into a groove milled in the edge of the deck sheet (see the bottom photo on the facing page). The most popular brand of bevel moldings is Wilsonart Custom

Custom beveled laminate edges eliminate the phenolic stripe and give the counter a solid-surface appearance. *Photo courtesy Wilsonart.*

Bevel-edge moldings with a spline-type tongue milled into the back hold better and provide more certain top-edge alignment than flat-back moldings designed to be taped on. A groove must be cut into the underlayment edge to accept tongued moldings. *Photo courtesy Wilsonart.*

Moldings, which used to be called Perma-Edge moldings before the company changed the name in 1995. The Wilsonart moldings are made of plastic laminate (Wilsonart styles only) bonded to a medium-density fiberboard (MDF) core with urea resin glue. Independent bevel edge makers will typically make edges using any brand of laminate. Some other brands of edge moldings are glued together with contact cement, and, though cheaper in price, they are not nearly as strong or as durable as moldings glued with a rigid adhesive like urea resin. Bevel-edge moldings are typically available in 12-ft. lengths.

One of the nice things about Wilsonart Custom Moldings is that the company also sells a complete power-tool equipment package for putting the edges on. What's not so nice is that the equipment package (see the photo below left) is expensive. When I started to write this book, I had never used bevel-edge moldings before, and all I knew about them was what I had gleaned from promotional product literature. I had a lot of questions about bevel edges and the Wilsonart tools in particular: How did the edges go on? How were the tools used? Were the tools absolutely necessary to do the job? Was the equipment package worth the investment?

I got the opportunity to find out the answers to those questions, because Wilsonart lent me a set of tools for a few months. I didn't get a chance to use them quite as much as I would have liked, but I was able to learn what I wanted to know. I found that the special equipment does make it quite easy to attach tongued-edge moldings, but, although I didn't try it, I have no doubt that you could do a decent job of installing the edges using regular woodworking power tools. If you are a professional who wants to add the laminate bevel to your repertoire of edge options, I think you will find that at least some of the Wilsonart edge tools are a worthwhile investment (components can be purchased individually).

In any event, here's an overview of how Wilsonart bevel edges are installed, using the Wilsonart tools. Whether you are interested in making one bevel-edged top for yourself or hundreds of

This collection of tools is sold by Wilsonart to help fabricators install its line of custom moldings. *Photo courtesy Wilsonart.*

them for eager customers in your area, it will help answer many of the basic questions you might have about bevel-edge fabrication in general.

A bevel edge is typically attached only to the ¾-in. edge of the deck, which means it will hang down in drop-edge form. After the edge is attached, the build-up is applied. A variation on this is to attach the front build-up so it's offset back from the edge, as shown in the drawing below); surprisingly, moldings with a glued-in tongue turn out to be strong enough on their own without needing backup support from the build-up.

After the top sheet of laminate is attached to the deck and trimmed flush to its underlayment, the edge that will be grooved to receive the molding must then be machined straight, smooth and at a perfect 90° angle to the top. A laminate trimming bit riding against the particleboard edge won't cut a smooth enough edge for getting a tight fit at the top of the bevel molding; you will need something better. The Wilsonart tool for this smoothing operation is the "straight-cut planer"—a router equipped with a special base and a straight-cutting router bit. If you made the countertop with straight and square-cut edges to begin with (as you always

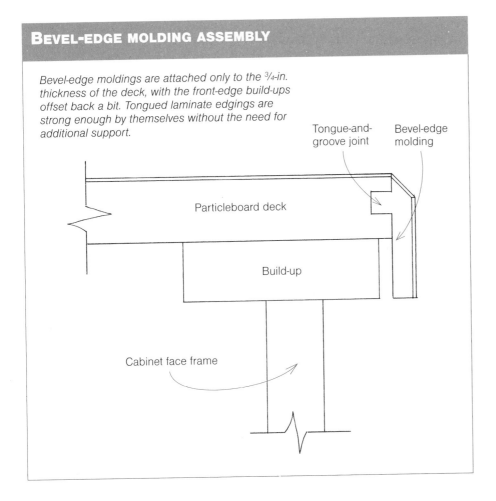

BEVEL-EDGE MOLDING ASSEMBLY

Bevel-edge moldings are attached only to the ¾-in. thickness of the deck, with the front-edge build-ups offset back a bit. Tongued laminate edgings are strong enough by themselves without the need for additional support.

Tongue-and-groove joint

Bevel-edge molding

Particleboard deck

Build-up

Cabinet face frame

should anyway), you might be able to do the necessary smoothing with careful use of a sanding block, sharp block plane or a power hand plane. If you have a good straightedge, you could use it in conjunction with a bearing-guided template bit to make a pass down the edges, and achieve the same result as the Wilsonart straight-cut planer. Adjust the straightedge and router to clean off about a 1/8-in. swath of edge.

The next step is cutting a 1/4-in. wide slot in the edge that will receive the tongue on the back of the molding. For this job Wilsonart provides another router with an oversize ("computer designed") base and a triple-wing bit with ball-bearing pilots positioned above and below the cutter. They call this tool the 1/4-in. dado cutter, though it actually cuts a slot and not a dado. The slot is 1/2 in. deep, which is plenty for the 1/4-in. tongue. Because the bevel edge must end up exactly flush with the top surface of the countertop, the slot must be located very precisely on the thickness of the deck. You set the alignment of the slot by adjusting the router's depth of cut, and then checking the fit on test pieces until it is just right. A regular router and base with the appropriate bit will also yield an acceptable slot.

The next stage is to cut to length and dry-fit the edge pieces. Although you could probably make all these cuts with a sharp carbide-tipped trim blade in a chopsaw, a disc sander with a miter gauge would prove very useful for fine-tuning the fit at outside and inside corners. The Wilsonart 10-in. disc sander

is, to me, the most impressive part of their equipment package (and the one tool I intend to get for sure). It is compact and powerful, fast, accurate, and easy to use. In addition, the sanding disc is mounted on an aluminum backer plate that does a great job at dissipating heat build-up. The sander comes with three factory-set fixed-angle miter gauges for making 90°, 45° and 22 1/2° miters. The tool also has a vacuum port to plug a shop vacuum or dust-collection system into.

Once all the edge pieces are cut and dry-fit, you glue the tongue, tap the pieces into position with a rubber mallet (a white mallet won't leave black marks) and clamp-tape them in place with filament tape. When the glue has dried, the sharp top edge will need to be ever so slightly smoothed down with fine sandpaper.

There is much more to learn about attaching laminate bevel edges, and you can get more detailed information by requesting how-to literature or calling Wilsonart's technical department (see Resources on pp. 127-129). Other companies that supply bevel-edge moldings will also be able to provide you with similar information and support regarding the use of their products.

The companies I know of that sell bevel edges also supply premilled wood moldings for use as flush-butted edging (see the photo on the facing page). When such moldings have a back tongue and are applied as I've just described, they will go on very nicely.

These Wilsonart moldings go together well and make for an easily fabricated flush-butted wood edge. The 2-in. radiused corner required some minor sanding and shaping to get a flush fit where it joined the straight edge pieces.

One thing you can't do with a bevel-edge laminate molding is make curved edges. However, there is a company, the Align-Rite Tool Co., that sells specialized tools for routing out and inlaying a bevel edge on a conventionally made square-edged top, and their equipment will make a bevel-edged radius (see Resources on pp. 127-129).

Another type of custom edge possibility recently introduced by Wilsonart is moldings made of Gibraltar, which is the company's solid-surface product. Strips of Gibraltar molding come as squared-off sections with a back tongue. They are applied similarly to laminate custom edges, but require a special glue. After installation, you can mill any detail you want into the edge with a router.

8

RESURFACING COUNTERTOPS

Countertops don't last forever. Most work surfaces will show the effects of everyday wear, or their pattern style will become outdated, before the cabinetry around them. Consequently, a great deal of my work as a fabricator involves making replacement tops. Although I usually prefer to build completely new countertops in these situations, it is possible to resurface a custom square-edge top, and there are times when I offer my customers the option. Specifically, I suggest resurfacing when it's clear to me that the existing countertop was well made to begin with and is still structurally sound. Built-in countertops with large or complicated layouts are the most likely candidates for resurfacing; post-formed tops can not be resurfaced.

The big advantage to resurfacing is that I can save myself the trouble, and my customer the expense, of ripping everything out and building a new underlayment. There are no disadvantages to resurfacing that I'm aware of; once completed, a resurfaced top is virtually indistinguishable from an original fabrication, and its functional lifespan will be no less.

You can resurface a countertop in two ways. A new layer of laminate can be applied directly over the old one (I call this laminate over laminate resurfacing), or over a piece of plywood, which is applied on top of the old counter (laminate over plywood resurfacing).

LAMINATE OVER LAMINATE RESURFACING

Laminate over laminate is a perfectly workable way to resurface a countertop, since laminate will adhere exceptionally well to another sheet of laminate. But this method can be used only if the original laminate-to-underlayment bond is still good. In the few instances where I've resurfaced directly over old laminate with a new sheet, the existing countertop was in excellent physical condition with absolutely no evidence of contact-adhesive bond failure. A sure way to determine the condition of the original bond is by tapping over the surface with your knuckles or with a hard object, such as the handle of a screwdriver; poorly bonded areas will make a distinctly different "loose" sound when they are tapped.

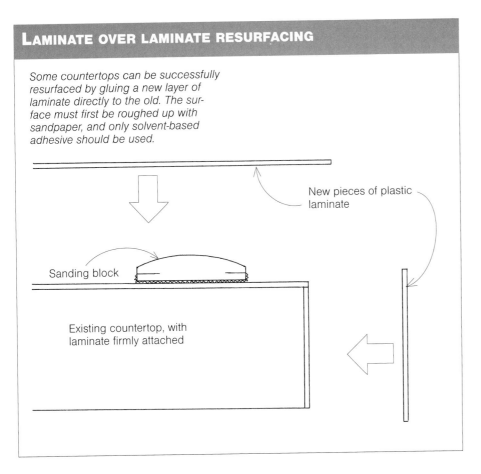

Some countertops can be successfully resurfaced by gluing a new layer of laminate directly to the old. The surface must first be roughed up with sandpaper, and only solvent-based adhesive should be used.

New pieces of plastic laminate

Sanding block

Existing countertop, with laminate firmly attached

To glue plastic laminate directly over plastic laminate, you must first thoroughly sand the entire top surface of the old countertop, not to remove the decorative paper but to roughen the glossy surface and provide tooth for the contact adhesive. To do this, I use a hand sanding block with 36-grit sandpaper (don't waste your time and effort with finer grits for this). An electric orbital sander will make the task easier, but, unless you hook it up to a vacuum, you'll be subjected to a lot more airborne dust.

After every square inch is sanded, clean up the loose dust, then wipe the surface down with a rag dampened with solvent (acetone) to get the surface thoroughly clean. Then apply the new layer of laminate as you would over a new particleboard underlayment (see pp. 73-83). However, be sure to use a solvent-based adhesive; water-based adhesives are not recommended for gluing laminate to laminate.

Another way to resurface old countertops is to glue and staple a skin of ¼-in. plywood over the existing laminate surfaces. This creates a new 'underlayment' for gluing the new laminate to.

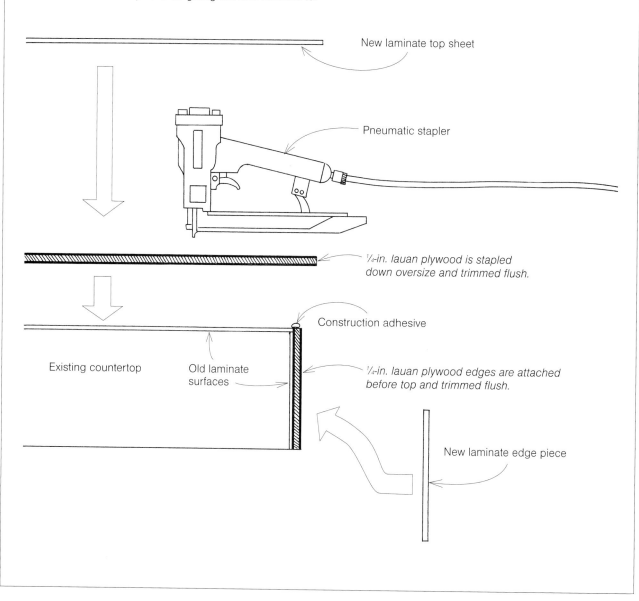

New laminate top sheet

Pneumatic stapler

¼-in. lauan plywood is stapled down oversize and trimmed flush.

Construction adhesive

Existing countertop

Old laminate surfaces

¼-in. lauan plywood edges are attached before top and trimmed flush.

New laminate edge piece

LAMINATE OVER PLYWOOD RESURFACING

The other way to resurface a countertop, and the one I rely on most often, is to fasten a layer of ¼-in. lauan plywood over the existing countertop and edges and then glue the new laminate sheets to that.

Select your plywood carefully—you want a good-quality plywood with a smooth face. The lumberyard I deal with carries a three-ply lauan that's put together with exterior glues, and I've found it to be a dependable product with a good gluing surface. However, I have on occasion also purchased ¼-in. lauan plywood made with interior glues from a discount home center, and, though it was cheaper in price, the veneer was loose in places and the surface was comparatively substandard.

Just as with a conventional particleboard underlayment, it is important to follow the Rule of 48 discussed on pp. 11-12; allow the plywood, laminate and adhesive adequate time to acclimate to temperature and humidity conditions in the fabrication area before assembly.

After removing any sinks and other counter inserts, you can put down the new piece of plywood. I apply the plywood to the edges first and then the top using glue and a pneumatic stapler. For glue I use general-purpose construction adhesive in a caulk cartridge and apply a small bead around the perimeter of each plywood piece. My pneumatic stapler shoots 1-in. long narrow crown (5⁄16-in. wide) staples, which are ideal because the heads countersink themselves and the legs splay as they're fired

in, thereby locking the fasteners in place. The depressions left by the staple heads are so tiny that they need not be filled. Don't be conservative with the staples; space them about 4 in. apart in a random pattern.

One of the nice things about the laminate-over-plywood approach to resurfacing is that the plywood pieces can be cut slightly oversize and the excess trimmed off with a bearing-guided, flush-trimming router bit in the same manner that you trim the plastic-laminate sheets (see pp. 90-91). Using this method makes the task of underlayment preparation very easy. Use body filler where needed, check the plywood surface carefully for staples that may not have set properly, hand-sand as needed to smooth down rough spots, and the underlayment is ready for laminate.

Since a countertop resurfaced with ¼-in. plywood will be that much thicker than the old one, you'll want to assess whether adding the extra layer will create any obstacles or problems. In this regard, the only hindrance I've ever encountered is rigid sink supply lines that ended up being ¼ in. too short for making a good connection. Typically, though, it makes good sense at this time to upgrade old drain connections and supply lines anyway; most customers also opt for a new sink and faucets too. The countertop edges will also be ¼ in. thicker. If the old counter edge was a standard 1½ in., the extra thickness will not be evident to the casual observer and does not look bad.

RESOURCES

The materials for making plastic-laminate countertops can be obtained at virtually any home center or building-supply yard. Such sources will usually have basic fabrication tools like carbide scorers, laminate files and standard trim bits. For the more specialized tools, there may be local sources, but you'll probably find it easier to rely on mail-order suppliers.

For technical information, the best source is the technical departments of laminate and contact-adhesive manufacturers.

The listings here are by no means exhaustive, but they will direct you to sources for laminate, adhesives, laminate trimmers and specialty tools—in short, every product mentioned in this book, as well as a few that I felt were too specialized for an introductory text.

LAMINATES

If you ever have a question or problem regarding any aspect of laminate fabrication (including the use of adhesives), don't hesitate to contact the manufacturer of the laminate. Virtually all of the companies have technical departments staffed with knowledgeable people. Over the years I've talked to many different technical representatives, and I've always found them to be friendly and helpful. Many companies also have technical sheets and topic-specific fabrication bulletins that they'll send upon request.

Formica Corp.
10155 Reading Rd.
Cincinnati, OH 45241
(800) 524-0159
(513) 786-3400

Interlam, Inc.
P.O. Box 22794
Ft. Lauderdale, FL 33335
(954) 581-5882
FAX (954) 581-3274
(Interlam is a distributor for Italian laminates.)

Nevamar
1278 Orgill Ave.
Memphis, TN 38106
(800) 526-9469
FAX: (901) 774-7277
(Nevamar laminates)

Pioneer Plastics Corp.
One Pionite Rd.
P.O. Box 1014
Auburn, ME 04211-1014
(800) 777-9113
(207) 784-9111
FAX (207) 784-0392
(Pionite laminate)

Ralph Wilson Plastics Co.
1061 NW H.K. Dodgen Loop
Temple, TX 76502
(800) 433-3222
(800) 792-6000 (in Texas)
(Wilsonart laminate)

ADHESIVES

Adhesive manufacturers are another reliable source of good technical and how-to information. There are manufacturers other than those listed here, but these are the ones I'm familiar with.

Columbia Cement Co.
159 Hanse Ave.
Freeport, NY 11520
(516) 623-6000
(Columbia manufactures adhesives for Formica Corp.)

Lokweld
(Contact Ralph Wilson Plastics Co., listed on p. 127 under laminate manufacturers.)

3M Industrial Tape and Specialties Division
3M Center Building 220-7E-01
St. Paul, MN 55144-1000
(800) 742-5933
FAX (800) 742-5933

V & S Sales Co.
12115 Burke St., Suites 4&5
Santa Fe Springs, CA 90670
(800) 669-8871
FAX (800) 554-8547

SPECIALTY TOOLS AND SUPPLIES

The basic tools needed to work plastic laminate (carbide scoring tools, files and trimming bits) should be readily available at any well-equipped home center. These items, as well as many other specialty laminate tools and related supplies, are also available from the two mail-order suppliers listed below.

JCM Industries
774 State Road 13, #9
Jacksonville, FL 32259
(800) 669-5519
FAX (800) 660-7371
(This mail-order company has the most complete selection of specialty tools and supplies for working plastic laminate that I know of, as well as specialty tools for cabinetmakers and solid-surface fabricators.)

Practical Products Co.
3925 Virginia Ave.
Cincinnati, OH 45227
(513) 561-6560
FAX (513) 561-6778
(Like JCM Industries, Practical Products offers a variety of specialty tools and supplies for mail-order shipment to laminate fabricators, though the product line is not as extensive.)

For specialty items mentioned in this book that are not offered by mail-order suppliers, you may need to contact the manufacturer directly for the name of a distributor near you.

Align-Rite Tool Co.
1942 East 17th St.
Tucson, AZ 85719
(602) 624-4438
(Align-rite makes a nice looking 60-in. long aluminum seaming jig called the Seam-Rite Fixture. The company's AR200S Bevel Edge System is the only one I know of that is capable of inlaying a bevel-edge strip on a radiused corner.)

Art Betterley Enterprises
P.O. Box 49518
Blaine, MN 55449
(612) 755-3425
FAX (612) 755-5084
(Art Betterley is a well-known pioneer in the development of special-purpose laminate power tools. The company's catalog and products are geared to the serious professional fabricator.)

Beno J. Gundlach Co.
211 North 21st St.
P.O. Box 544
Belleville, IL 62222
(618) 233-1781
FAX (618) 233-3636
(Wholesale supplier of many hard-to-find hand tools for working laminate, tile, vinyl flooring and carpeting)

D&K Sales
12972 Farmington Rd.
Lavonia, MI 48150
(800) 637-2250
(Distributor of Smart Clips for backsplashes)

DeWalt Tools
701 E. Joppa Rd.
Towson, MD 21286
(800) 433-9258
(Manufacturer of high-quality laminate trimmers)

Kampel Enterprises
8930 Carlisle Rd.
Wellsville, PA 17365
(717) 432-9688
FAX (717) 432-5601
(Manufacturer of colored filler for laminate repair)

Kett Tool Co.
5055 Madison Rd.
Cincinnati, OH 45227
(513) 271-0333
FAX (513) 271-5318
(Manufacturer of electric laminate shears)

Klenk Industries
20 Germay St.
Wilmington, DE 19804
(800) 327-5619
(Manufacturer of laminate hand shears)

McFeely's Square Drive Screws
1620 Wythe Rd.
P.O. Box 3
Lynchburg, VA 24505-0003
(800) 443-7937
FAX (800) 847-7136

Porter-Cable Corp.
Youngs Crossing at Hwy 45
P.O. Box 2468
Jackson, TN 38302-2468
(800) 487-8665
(Manufacturer of high-quality laminate trimmers)

S-B Tools (Bosch)
4300 West Peterson Ave.
Chicago, IL 60646
(312) 286-7330
(Manufacturer of high-quality laminate trimmers)

Severance Tool Industries
P.O. Box 1886
3790 Orange St.
Saginaw, MI 48605
(517) 777-5500
(Manufacturer of carbide files)

Simp'l Products
264 Fordham Place
Box 187
City Island, NY 10464
(718) 885-3314
(Manufacturer of the Laminatrol Cutting Guide)

WoodsmithShop
2200 Grand Ave.
P.O. Box 842
Des Moines, IA 50312
(800) 444-7002
FAX (515) 283-0447
(Mail-order supplier of T-tracks and knobs for making your own seaming jig)

WOOD AND LAMINATE EDGE MOLDINGS

Kuehn Bevel
111 Canfield Ave.
Randolph, NJ 07869
(800) 862-3835
FAX (201) 584-1855

Wilsonart Custom Moldings
(Contact Ralph Wilson Plastics Co., listed on p. 127 under laminate manufacturers.)

INDEX

PUBLISHER: **Jon Miller**

ACQUISITIONS EDITOR: **Julie Trelstad**

EDITORIAL ASSISTANT: **Karen Liljedahl**

EDITOR: **Ruth Dobsevage**

DESIGNER: **Henry Roth**

LAYOUT ARTIST: **Amy L. Bernard**

PHOTOGRAPHER, EXCEPT WHERE NOTED: **Sloan Howard**

ILLUSTRATOR: **Herrick Kimball**

TYPEFACE: **Frutiger Light**

PAPER: **70-lb. Moistrite Matte**

PRINTER: **Quebecor Printing/Hawkins, New Canton, Tennessee**